A Legacy of Promises

The Texas Pan American Series

For Pastora, & Santiago and Pablo

Guillermo de la Peña

A LEGACY OF PROMISES

Agriculture, politics, and ritual in the Morelos highlands of Mexico

University of Texas Press, Austin

The Texas Pan American Series is published
with the assistance of a revolving publication
fund established by the Pan American
Sulphur Company.

International Standard Book Number 0–292–74630–X
Library of Congress Catalog Card Number 81–50486

Printed in the United States of America

CONTENTS

LIST OF TABLES AND FIGURES

Tables

Figures

ACKNOWLEDGEMENTS

I owe an intellectual debt to Angel Palerm, Bryan Roberts, Norman Long, Bruce Kapferer, Martin Southwold and Roberto Varela. Angel Palerm introduced me to the subject of anthropology and was present practically throughout all the stages of this book. The others gave generously of their time and ideas; their comments and criticisms have been a major stimulus for me. Of course, my debt to all six is not only intellectual; they are very good friends.

While living in Tlayacapan, I enjoyed the friendship and generosity of Claudio Favier, Paloma Calderón, Javier Cabrera, Angel and Guadalupe Rojas, Pedro and Irene Bailón, Jaime Rojas, Gustavo Amaro, Jesús Olmos, Primo Vidal, and many other people whom it is impossible to list. Salvador Núñez assisted me greatly in the study of local religion.

In the fieldwork I received the invaluable help of several students: Cecilia Lezama, Lilliam Barrios, Pilar Charles, Angeles Sánchez, Diana Molina, Lourdes Herrasti, Julia Sierra, Luisa Medina-Mora, Laura Helguera, Patricia Santos, María Sánchez, Marta Vidal and Graciela Caraza. Carmen Campero, Marta Berea and Ana González helped in the documentary research. Many of the colonial documents I mention were located and copied by Teresa Rojas; I am very grateful to her for having allowed me to make use of them.

I benefited from comments and criticism in the staff and postgraduate seminars at the University of Manchester. Particularly, I should like to thank Elinor Kelly, Michael Gilsenan and Ken Brown. The latter and Jorge Dandler made pertinent suggestions during my doctoral examination.

David Brading and Sue Hogg made valuable criticisms when I was editing the manuscript. Several institutions have given me financial support throughout different stages of the research: the Ford Foundation, the Centro de Estudios Educativos, the Centro de Investigaciones Superiores del Instituto Nacional de Antropología e Historia, and the Universidad Autónoma Metropolitana.

The last period of writing could not have been completed without the support and hospitality of Susan Roberts and Sarah Southwold.

I am grateful to Ruby García, Rajinder and Jennifer Vaughan for their efficient typing, and to Bush Kelly for drawing the maps.

Lastly and most important, the constant encouragement of my wife Pastora sustained my interest in, and commitment to, this book.

All these people contributed to the possible merits of this work; its multiple errors are exclusively mine.

I
INTRODUCTION

> Communities should be viewed no longer as self-contained and integrated systems in their own right. It is more appropriate to view them as the local termini of a web of group relations which extend through intermediate levels from the level of the community to that of society. [Eric R. Wolf, 1965:85]

Heterogeneity and interdependence

It is hardly original to say that Mexico is a heterogeneous country. Its territory presents a variety of climatic zones and geological formations: tropical valleys and snow peaks, deserts and forests, huge mountain ranges and open plains. Similarly, the Mexican population is made up of widely contrasting social categories and groups. Their occupations are different; as are their material standard of living, their relations to the means of production, and their access to resources in general. In addition, it is frequent to find neighbouring *pueblos*[1] which present styles of living—customs, symbols—markedly different.

To explain this heterogeneity, anthropologists have drawn attention to certain characteristics within groups, sectors or human settlements. Some have pointed to the impact of geographical factors in determining discrete ecological organisations (Sanders, 1967). Others have referred to separate 'traditions', or internal processes of cultural transmission, whereby different symbolic and behavioural patterns emerge and persist. Others speak of higher or lower degrees of acculturation—depending on the intensity of contacts between 'natives' and bearers of European 'western' cultures (Beals, 1967), and on the power relationships implied by these contacts (Aguirre-Beltrán, 1967, 1970). In this book I stress the importance of other types of factors related to heterogeneity. I follow the hypothesis that the plurality of human settlements and groups is above all a function of their interdependence within a process of national, vertically organised consolidation. I believe this approach is necesary to complement previous studies which have focused their attention on the peculiarities of a single village, area or ethnic group.

There are nowadays many studies which emphasise the necessity of placing in a wider context the small observable universe which the anthropologist investigates directly.[2] However, community studies still form the bulk of anthropological output in Mexico. Sometimes the insistence on the community approach leads to obvious distortion, in so far as it fails to recognise that the organisation of social life in a community responds to forces which transcend local boundaries.[3]

It is paradoxical that both the community approach and the trend to studying wider social environments have a heavy intellectual debt to the same man: Eric R. Wolf. In two masterful essays (1953, 1965), Wolf demonstrated the crucial importance of inter-group relationships and nation-building processes for the understanding of even a single aspect or part of a complex society like Mexico. He was also the first anthropologist who achieved a synthetic view of Mexican cultural and social processes, in his book *Sons of the Shaking Earth* (1959). However, Wolf's influential work on the definition of the community—'supported in its autonomy by a grant of land, charged with the autonomous enforcement of social control [each community] constituted a small, closely defended *island*,[4] securing the social and cultural homogeneity of its members within, struggling to maintain its integrity in the face of attacks from without' (Wolf, 1967: 299)—encouraged the development of monographs which paid special attention to social relations within a settlement and minimised the insertion of individuals and groups in wider processes.[5]

Wolf himself stressed the necessity of looking at the community as a unit having links with external institutions such as *haciendas* or *encomiendas*.[6] Yet even in his writings the community remains as 'a kind of monad whose internal dynamics and whose trajectory cannot be grasped more than obscurely' (Leeds, 1965: 399). Therefore, the nature and dynamics of external links are also difficult to grasp. Further, his idea of the community having an 'autonomous', persistent organisation is empirically validated only by general and rather vague reference to the literature. The *a priori* quality of the community concept relates Wolf's insights to Redfield's conceptualisation of the stable, 'folk', timeless, harmonic 'little community' (1960, 1965), where conflicts and change are explained as recent pollutions brought from the outside[7]—although the real past of the communities is seldom properly investigated. Although the 'folk' conception echoes Durkheim's ideas, its application has often implied a disregard for Durkheim's rule: 'les phénomènes sociaux doivent varier, non pas seulement suivant la nature des éléments composants, mais suivant leur mode de composition . . .' (1968: 85).

Modernisation, unilineal change and the peasantry

A corollary of the emphasis on interdependence is that the process of national consolidation is not equivalent to a process of homogenisation within an inclusively capitalist economy. In other words, the existence of, for example, peasant groups must not be seen as a sort of *defect* in a nation—an obstacle which should be overcome in order to achieve 'modernity'. A leading exponent of so-called 'modernisation theory' has written, 'Although the rudiments of capitalist (i.e. market) economy are present and important in peasant communities, they are *incomplete* and *underdeveloped* compared to market organisation in a modern national economy' (Dalton, 1971: 209–10;[8] *his italics*)

The view advanced here is that, in the case of Mexico, the national economy—more precisely, the process of capital accumulation—has entailed the existence of 'non-modern' sectors, articulated to 'modern' organisations. Furthermore, each one needs the other to survive and 'develop'. Peasantry does not precede capitalist agriculture, or industry, in a unilineal evolutionary scale; both march together in a sort of love–hate relationship (Warman, 1976: 293 ff.).

In another article Dalton claims that the process of modernisation is essentially the same in all parts of the world (1972: 405–6). He virtually repeats some of Redfield's arguments, saying that the emergence of national institutions in industrial societies 'displace' local, previously isolated institutions. This seems to imply that macro-social institutions reaching down to the village level are the product of contemporary industrial capitalism—which is historically false.[9] In this connection, Eric Hobsbawm has rebuked modernisation theories, arguing that they 'are anti-historical in two ways: in substituting the empty progress from "traditional" to "modern" for what happens to real societies, and in reading the past . . . essentially as the preparation for a satisfactory, though not yet perfect, present' (1975: 27).

Mutatis mutandis, there are some points of contact between unilineal theories of modernisation and development, on the one hand, and certain orthodox Marxist formulations, on the other. An example of the application of the latter to Mexico is Roger Bartra's writings. He contends that the peasantry tends to become either rural proletarians or—in a much lesser proportion—capitalist farmers (1974: 78–9), i.e. that agricultural production as a whole marches *towards* the organisational form of modern capitalistic firms. (He does not document how this process is taking place.) This approach has its origin in the analyses by both Kautsky (1974) and Lenin (1976) of European and Russian agriculture at the turn of the century, where the disappearance—and destruction—of the peasantry ('that paradise of petty-bourgeois, conservative writers': Lenin, 1976: 12) is seen as a symptom of progress, in so far as it relates to technical advance and also to the growth of a progressive class: the proletariat. (It is argued that peasants are not 'progressive' but fanatical and individualistic.) Kautsky stressed the positive impact of modern machinery on rural life: it pushes workers into the cities (thus fostering industrialisation) and raises rural incomes (1974: 45 ff). Lenin even praises the results of usury and mortgages because they help to separate the peasant from his plot and reinforce the expansion of large-scale enterprises—which are able to minimise costs, maximise results and, through accumulation of technical knowledge, improve the quality of the labour force (1976: 12, 17, 18–19, 24–5).[10]

It seems less easy to accept these propositions for the case of Mexico. The countryside stubbornly refuses to become fully organised in terms of modern firms. In turn, urban industrial enterprises have never been capable of absorbing more than a small proportion of the national labour force (Palerm, 1968). The trend towards rural mechanisation—and the concomitant redundancy among the rural labour force—fosters not industrialisation but poverty, both in the country and in the city. On the other hand, the rural population itself frequently opposes—and in fact limits—'modern' trends. One crucial cause of the 1910 revolution in Central Mexico was the opposition to the expansion of big enterprises—i.e. the unsolvable contradictions created by such expansion.

On the definition of peasantry

The discussion of interdependence and modernisation is linked to a rather entangled disagreement about the meaning of the word 'peasant'. There seems to be a certain consensus that peasants are small-scale rural cultivators, that they have a certain control over land, and that they are somehow connected to a larger society (Kroeber, 1948; Redfield, 1953). In consequence, it is also agreed that peasants are *not* tribesmen (rural groups whose economy is directed towards subsistence and who are not structurally part of a complex heterogeneous society), nor large-scale agricultural entrepreneurs, nor rural workers who receive a salary and have no direct control over land. However, different writers present different versions of how peasants organise their lives, and how they relate to the larger society:

1. Some see peasants as 'transitional': they are destined to be totally devoured by the large-scale organisational mechanisms of the nation. Their present organisation is therefore a mere initial stage in a process of integration. This view of the peasantry is germane to both the modernisation and the unilineal Marxist approaches already mentioned.

2. Others stress the resilience of peasants to the invasion of the outside world: in spite of their contact with the outside market, and generally with cities and the wider society (through formal institutions or informal agents), their primary organisational units are still the household and the local community (Fallers, 1961). In this sense it is possible to speak of peasant social, economic and political life (Redfield, 1960).

One could further subdivide this approach according to the stress given to sociological or psychological factors in studying the internal consistency of peasant life. Thus Manning Nash attributes to peasants 'the possession of a set of concrete social organisations which directly channel economic choice, on the one hand, and a set of sanctions which operate to keep economic deviants in physical as well as moral jeopardy on the other' (1968: 311). Similarly, Frank Cancian (1965) sees the organisation of *fiestas* in rural Mexico as the structure sealing off the cultural space in which conduct occurs, and as the main explanatory variable for social life. In their turn, George Foster (1960–61, 1967 a, 6), Michael Maccoby (1967), John Ingham (1970) and others postulate the existence of peasant cognitive orientations or psychological configurations, which determine people's perception and therefore behaviour and resistence to change.

3. An influential Soviet scholar (later banned in his own country), A. V. Chayanov, stressed the peculiar economic orientation of the peasant household, and argued that it responded to specific economic laws. The crucial variable in understanding these laws is the double dimension of the household as a production–consumption unit. The lowest limit of household production is therefore determined by its minimum consumption needs;[11] its highest limit, by the capacity of the family labour. The goal of peasant production is primarily family subsistence; this also determines the work orientation of individuals. In consequence,

even if individuals participate in a non-peasant labour market, they will do so as a supplement to consumption needs, and their organisational framework will remain primarily that of the production–consumption household unit (Chayanov, 1966).

4. A fourth approach holds that the peasantry is a result of the impact of wider society—especially capitalist society. This approach has roots in Marxist studies of colonialism, like those by Paul Baran (1957) and—earlier—Rosa Luxembourg, which draw attention to the fact that pauperisation of 'backward' areas is a function of their contact with the capitalist world, in so far as the latter extracts raw materials and cheap labour from them (Stavenhagen, 1968: 16). The leading exponent of these ideas is probably André Gunder Frank (1970, 1971), who has repeatedly contended that capitalism has been present in Latin America as the main organising force since the sixteenth century.

5. A fifth approach focuses on the asymmetrical nature of the links connecting the peasantry with the wider society. It differs from Frank's ideas in emphasising that the peasantry can be linked to various types of 'social segments':[12] despotic bureaucracies in 'Asiatic', tributary societies; landlords in feudal States; capitalist agricultural entrepreneurs and government agencies in contemporary nations (Wolf, 1966; Shanin, 1971; Warman, 1973). It is argued that the working principle behind the forces integrating rural settlements with wider societies is as old as the emergence of urban civilisations and rudimentary nation States: it arises from the systems of appropriation of primary wealth by a population of specialists who are not directly engaged in producing food (Wolf, 1966: 10). This point of view also maintains that the different segments do not obey the same evolutionary laws, and that their internal dynamics should be analysed separately. For these writers, therefore, the peasantry can be studied in its own right, whereas for Frank the capitalist system is the sole object of study.

6. Yet another definition of peasantry should be mentioned: that which sees it as a 'mode of production' *articulated* to other 'modes of production', specifically to the capitalist one. This position has many similarities to the 'segment' or 'multi-evolutionist' approach—in fact writers such as Warman (1976) use either terminology indiscriminately—but the internal characteristics of peasant organisation are more carefully spelled out in terms of 'forces of production' and 'relations of production'. Some authorities prefer not to speak of a peasant mode of production, but rather of a 'petty-commodity' one, so that activities such as fishing and handicrafts could be included (Amin, 1975). What

characterises this mode of production is, first, its inability to generate more capital than that necessary simply to perpetuate itself (i.e. there is no proper growth within the mode); second, the overlapping of its relations of production with those of other modes, usually dominant. For instance, the use and disposal of produce is mediated in contemporary peasant societies by the disadvantageous participation of peasants in capitalist markets.

It is impossible properly to discuss even one of these approaches within the limited space of this introduction. On the one hand, they show that, in the literature, the term 'peasant' is far from having an unequivocal meaning. In fact the indiscriminate and superficial use of the term has been severely criticised (Ortiz, 1971). On the other hand, not all the approaches are mutually exclusive. For instance, Roger Bartra espouses the 'transitional' notion of peasantry, while at the same time accepting that the also transitional petty-commodity mode of production does articulate with the overall capitalist system (1975). The ideas of Chayanov have been used both by authors close to Redfield, such as Nash, and by neo-evolutionists like Wolf and Warman. And so on.

In spite of their heuristic utility, all these approaches raise some questions and pose some problems which have not been clearly answered or elucidated.

On the one hand, the particularistic approaches to peasant social structures fail to analyse their integration into the wider society. They usually fall into a sort of circular argument: peasants are linked to the wider society because by definition they are connected outwards. The specific nature of the links is seldom empirically established except in terms of unidirectional diffusion—and then the whole issue of heterogeneity is reduced to an ahistorical definition of transitional stages. In addition, one has to ask whether particularistic theories really help us to understand the behaviour of the people categorised as peasants. For example, precise hypotheses such as Chayanov's are probably useful in a situation of high stability, but they tend to be less fruitful when peasants are deeply involved in cash-crop production—which is primarily organised by market demands and control of inputs such as fertiliser, and not by household labour; when wage labour is the norm rather than the exception; when land is not available to expand subsistence production; and when migratory processes split the unity of the household (that is, Chayanov's hypotheses are hardly applicable to Mexico's situation of accelerated change). To put it simply, when the focus of study is just *the peasantry*, it is impossible to explain—or even to describe—the dialectics of observable social processes in most rural settlements, which are not reducible to the fulfilling of norms imposed by local ('peasant') institutions, nor to the meeting of demands made by household or community relationships. Demands and pressures made by outsiders—individuals, groups, enterprises—and the institutional norms of the wider society, which are often diversified and contradictory, are strong determinants of local behaviour; in turn, the latter modifies and conditions external influences.

On the other hand, approaches which stress the general, vertical influence of capitalist society on the peasantry refrain from exploring or explaining the variability of this influence at the local level. The problem of regional heterogeneity is thus left on one side; as is the phenomenon of the internal differentiation of peasantry. To say that heterogeneity and differentiation are unimportant is, theoretically, to beg the question and, analytically, to give up the investigation of the uneven impact of capitalism on the countryside and of the complex patterns of interaction and feedback between country and city.

For operational purposes, there is a lack of heuristic tools for the analysis of the empirical universes encountered by anthropologists: not peasant 'segments' or 'modes of production', but groups of interacting people who are far from homogeneous. In these universes, social action results from the processes of changing alliances among various categories of people. Action cannot be qualified merely as 'peasant' or 'non-peasant'. And if this is the case, then it is impossible to refer the analysis to a single structure, or to a set of clearly defined structures interconnected by clearly defined links. It is necessary to assume a context of multidimensional complexity.

The terms *social field* and *power domain* seem useful for approaching such a context. The problem of the definition of the peasantry—which in the case of Mexico is also the problem of the role of non-capitalist organisations in the process of capital formation—will be re-examined at the end of the book.

Social fields and power domains

In the anthropological vocabulary 'field' has become a standard term which is used analytically to deal with complex universes. When studying modern Zululand, Max Gluckman (1958) confronted many situations where the application of tribal law or customary norms was irrelevant and where it was likewise impossible to understand those situations solely in terms of the rules of the European colonialists. What existed was a new context of choice, in which individuals manipulated material resources and several normative paradigms according to their particular interests and perceptions. A social relation in this context had to be seen as *emergent*, i.e. 'as the joint product of [all] parties to it . . . [since] the action of one partner cannot be understood without reference to the action of the other[s]' (Kapferer, 1972: 6–7). A given set of social relations emerging in a given context received the name of *social field*.

The concept was subsequently used in studies of complex societies. These studies[13] helped to establish some of the properties of a social field: (1) It was usually heterogeneous, not only because the actors had divergent characteristics, but also because norms, symbols and perceptions varied. (2) It was multidimensional, because it could not be reduced to any single institution or social aspect (politics, kinship, religion, economics). This was an important innovation in anthropological discipline, which was often conceived as 'the study of social institutions' by the followers of Malinowski and Radcliffe-Brown. 'Social structures', in the sense of enduring social relations, should be taken into account in a field analysis, but are never coterminous with it. (3) It was not 'fixed' in terms of its components, or in terms of time and space. A field could expand and contract as actors and resources become relevant or irrelevant for the analysis.

The operational criterion of relevance was usually established in relation to a particular dynamic aspect of social life—i.e. a process, which could be observed through its different stages. The concept of *social forces*—used by some leading political anthropologists (Swartz, Turner and Tuden, 1966; Swartz, 1968; Turner, 1974)—isolates the essential elements of a process: those pressures put by participants on one another's behaviour. On the other hand, this kind of analysis inevitably flows into the issues of conflict and power.

It is implied in Durkheimian solidarity models and in Parsonian systemic theories that society is ultimately integrated by a convergence of interests towards shared goals, which results in the elimination of discrepancies and opposition through the implementation of redressive social mechanisms. But studies such as Evans-Pritchard's *The Nuer* (1940) demonstrated the existence of opposition as an essential aspect of social life, even of tribal (supposedly 'harmonic') social life. In some cases, these oppositions could be successfully integrated into the system. Conflict was never eliminated, but it did not lead to a disintegration of society, nor to basic, structural changes. Gluckman (1955) elaborated on Evans-Pritchard's arguments and coined the term *repetitive change* for those processes which implied continuously emerging conflicts within a persistent overall social structure. He argued that discrepant rules could be dealt with through a process of *situational selection*[14] —whereby individuals could interpret and manipulate normative frameworks according to actual circumstances—and through the multiplicity of social affiliations and cross-cutting ties: the fact that most individuals who were opposed in reference to one instance were united in reference to many other instances. However, Gluckman also held that not all societies presented processes which could be analysed in terms of repetitive change. The opposition of interests existing, for instance, in colonial South Africa, or in colonial Rhodesia, should inevitably lead to structural changes. Thus the study of social fields—which, by definition, entail multiple discrepancies and oppositions—usually implies the analysis of processes of structural change (not in one, but in several structures). Situational selection and cross-cutting alliances are still present in such processes, but they do not prevent the eruption of major crises, which can be solved only by a rearrangement of relations which previously had an enduring quality. On the other hand, the nature and direction of structural change is not random: it is conditioned by power.

One of the major contributions of sociological exchange theory has been to clarify the meaning of power by defining it as an emergent property of persisting imbalanced exchanges among given actors. Peter Blau (1964: *passim*) draws attention to the importance of control over those resources which are positively valued by a set of social actors; the actor(s) who exerts this type of control is in a position of dominance over other actors provided the latter accept their dependency on the controlling party by obtaining valued resources in exchange for their compliance. However, a power relation is not something static, but develops and changes as actors adopt different strategies in respect to the use of resources. An actor who at a given point of time does not exert control over a given resource could avoid compliance by renouncing the use of that resource, or by taking it by force, or by looking for alternative ways of gaining access to it (Blau, 1964: 118 ff.). He could also negotiate by offering another resource in return; a power relation is neutralised if the accepted evaluation of the resource given in return is such that the resulting exchange can be defined as balanced by the actors involved.

On the other hand, even if *all* power relations are being continuously redefined by strategic action, they vary according to their degree of inclusiveness and endurance. Richard Adams has used the word *environment* to transcend the rather atomistic view of power based on control over discrete resources. The concept of environment points to the whole set of resources that a given social unit needs in order to operate. These operational sets of resources are patterned in the case of persisting social units. For instance, an agricultural enterprise cannot operate as such without land, water, labour, technology and access to a market. All these resources are needed in a certain amount and in a certain order. Whoever gets control over this patterned environment is in a position to exercise power over the enterprise.

Adams develops his argument by suggesting that power relations are usually exercised not through physical coercion but through the recognition of the tactical control of the parties. As in Blau's discussion, this means that there are shared evaluative criteria among a given set of actors; but Adams is also interested in discovering a structure of both evaluation and control. For this he introduces the idea of *domain*, which 'exists when one party has greater control over the environment of a second than does the second over the environment of the first' (1970: 119). Different forms of domains may provide clues to the different mechanics of power relations. For instance, there are domains within domains, which entail the existence of derivative power (as opposed to independent power). A multiple (as opposed to a unitary) domain—where several parties control different aspects of the environment of a single subordinate party—entails fragmented power. A domain which is both unitary and independent is what comes closest to absolute power; but this still does not say anything about its extension (or range), since it could refer to a single subordinate party. If the same party holds a superior position in several domains its power will be extensive, but the *intensity* of power will always depend on the degree of actual tactical control which is being exercised over an environment.[15]

These concepts—imbalanced exchanges, strategies, environment, tactical control, domains—help to throw light on the previous discussion of social fields and structural change. First, power is an essential component of any emergent social field, i.e. the choices of the actors are limited by constraints derived from power relations. Second—and consequently—changes in structural relationships may be fostered or prevented through the exercise of power. For instance, a series of crises brought about by technological innovation in a rural area could result in changes in the structure of the family, but not in the structure of property, if there are actors who have the will and the power to prevent that outcome. In other words, there may be constraints which prevent such actions from leading to certain structural changes. Third, the actors who participate in a social field may or may not be co-participants in one or several domains. For instance, one of the actors may be a superior party to an independent, unitary, extensive domain in which all other actors are inferior parties. In such a case the latter would have a context of choice limited by a common frame.

I am interested in the common limiting frame of action provided by a particular domain: the State. The State (a capital letter, to distinguish it from the states or provinces within the federal Mexican republic) has been approached in the literature from two main perspectives: the Weberian and the Marxist. The Weberian perspective views the State as an empirically observed entity, characterised by its monopoly over the legitimate use of physical coercion (Weber, 1946: 78 ff.). The Marxist approach is more abstract: it views the State as a set of power relations and power mechanisms which are directed to the maintenance of a given class system. Hence this approach includes within the concept of the State not only the government apparatus but ideological devices and generally all elements which reinforce the hegemony of the ruling class and the perpetuation of the class system (Althusser, 1973).

The Marxist approach allows the researcher to focus attention on a variety of social relations, the functioning of which is essential for the successful implementation of government policies. I suggest that this perspective can be operationalised by seeing the State as a trend towards the implementation of a single, unitary, independent, extensive and intensive power domain over a territory. This permits one to do research on the empirical manifestations of the State in relation to the environments of relevant acting social units, as well as on the multiple oppositions and contradictions emerging in the process of consolidation of State power.

The argument

The subject of this book is the use and organisation of land and labour in a 'peasant' region called the Morelos Highlands. The use and organisation of land and labour are seen as a process in which changing conditions are taken into account and manipulated by relevant actors.

These actors are not only those living in the region, but many others who are 'external' in terms of the regional spatial setting. In addition, 'internal' actors are seen as conditioned in their strategies by their own relationships with different types of 'external' individuals and groups. Thus the specific patterns of land and labour which are to be found in the Morelos Highlands have to be understood in the context of the region's relations with other areas. A special emphasis will be placed on the analysis of the relations between the Highlands and the lowland region near by, because it is in the context of these relations that many relevant elements of both areas become intelligible. In addition, attention will be paid to the role of the State throughout different periods of time in so far as it provides a conditioning frame for the use of resources and for the emergence of basic social categories. The study has more historical depth than is often used in anthropological studies. Initially I planned to limit the analysis to 1970–71—the period in which I did more intensive fieldwork. In these years the effects of factors such as involvement in cash-crop export production, the presence of federal government agencies, and penetration by urban manufacturers were noticeable, as were the oppositions and maladjustments brought in their train. However, it was very difficult to identify and analyse processes of change within such a short time span. In the first place, the life histories and oral traditions I collected indicated that social change was not at all a recent phenomenon. People seemed to have been always involved in all sorts of affairs—from labour migration to itinerant commerce, from cattle thieving to revolution. Second, the actual organisation of local farming appeared to be related to a tradition of dry-season employment in the irrigated lowlands. This tradition was institutionalised in the past; thus an important insight into what was happening in the Highlands would be gained by focusing attention on the points of articulation between this region and the lowland sugar-cane *haciendas* during both the nineteenth century and the colonial period. This also helped me to understand more clearly the local emergence and effects of the 1910–20 revolution, which is both the event that people conceptualise as the most crucial in local history (after all, Morelos is the land of Emiliano Zapata) and the source of today's rural organisation.

I decided to devote two full chapters to the analysis of some historical data. I say *some* data, because a proper historical coverage of even a limited period of the past of Morelos would need long-term, exhaustive documentary research. My main interest is to show that some important characteristics of the Highland villages emerge from a historical process in which local land and labour resources have been combined within the framework of lowland agricultural organisation; also, that local crises and conflicts have to be explained as stages in this same historical process. The Highlands have for centuries been a reservoir of grain and—most important—seasonal labour for plantations in the lowlands. An important local function of the State has thus been to create and reproduce this pattern. The pattern demands that the Highland people have access to agricultural plots to grow their own food during the rainy season—when plantation jobs are scarce—and that they should be available in the dry season. This availability in turn depends on the lack of attractive dry-season alternatives for the Highland people. On the other hand, dry-season employment should certainly be available, otherwise the economy of Highlanders would suffer severely. Local crises have emerged when Highland people have not had enough land to grow food during the rainy season—as happened before the 1910 revolution—or when there is not enough dry-season employment—as has been the case in the last few decades.

In addition, historical information helped me to become aware of the inadequacies of, for example, the 'modernisation' type of approach for analysing a region like the Morelos Highlands. In the light of history, none of the suppositions of such an approach stands: the villages were never isolated; social differentiation has always been present; people were not mere executors of local norms, but participated in a wide social field and manipulated heterogeneous normative frameworks; changes were not the result of the impact of external forces, but of processes where local actors actively participated. In this sense, the analysis of the past provided an important perspective for the analysis of the present.

Chapters V to VIII are devoted to contemporary data. A point of departure is to consider local participation in, and some crucial effects of, the revolution. The land was distributed among villagers. The large sugar *haciendas* disappeared. Lowland suger-cane production decreased. The contraction of sugar-cane employment during the dry season is related to the Highlanders' involvement in other sorts of activities directed to the acquisition of cash: long-term migration, non-agricultural local activities, education as a means of getting urban jobs, and notably the growing of export cash-crops in land which had previously been used for the production of local food. All these activities, and principally the last one, have in turn led to important changes. One of them is in local labour organisation: given the requirements of big labour inputs for the growing of such new cash-crops as tomatoes, agricultural tasks—which previously could be done by members of the household—now demand the hiring of labourers. This demand for labour has meant that many local people who had migrated have come back to the Highlands, and it has also attracted people from other areas. The local population has grown, and this growth has three consequences: the competition for dry-season labour increases; hence there is a vital need to use the local land for cash production; there is more pressure on local land and more competition for it. Cash-crop production has also meant the introduction of technological innovations which favour increasing yields: improved seeds, chemical fertilisers and insecticides. This technology costs money. Hired labour costs money. Access to cash is a requisite for cash-crop farming. The heavy debts which many farmers incur often cannot be paid because of the risky nature of the specific crops they have chosen to grow, and because small-holding, cash-crop farming is inevitably risky. Many people sell their land in order to pay their debts. This fact, together with differential participation of people in other activities, is leading to the emergence of differentiated social categories defined primarily in terms of access to cash and resources.

Of course, all these contingencies are not simply caused by the decrease of dry-season employment in the lowlands, though initially this was a crucial factor. Other factors have to be examined as well. The demographic growth has a more complex origin than mere attraction of labour to local cash-crops; for instance, it is related to family patterns and government action in the field of public health and sanitation. The need for cash is related to the import of urban manufactures into the region through the emergence of new commercial networks. In turn, this is related to the government-sponsored industrial boom of 1940–60, and to the expansion of government-built communication services. The government is a crucial actor in contemporary processes. Among other things, its presence helps to explain why agricultural land in the Highlands—which has a market value, and is in fact coming on to the market—has not been monopolised by large-scale agricultural firms. The answer is to a large extent related to government commitment to a programme of agrarian reform, and to government development policies which favour small-scale, flexible enterprises.

To examine government action is to enquire into the presence and operation of the State, which I have defined as a trend towards the consolidation of an overall power domain over a territory. The theme of the State runs throughout the whole book. In relation to it, the historical approach helps to make useful comparisons between three distinguishable stages of State formation in Mexico. The colonial State was able to consolidate an overall domain through the implementation of detailed centralised mechanisms for the division of labour and the use of resources. The Church provided an important support for the maintenance of the domain. Religious ideologies put a seal of approval on the existing order; in addition, the organisation of ritual at the local level was closely interwoven with the mechanisms of labour control and resource allocation. After independence, this overall domain collapsed; there was no proper national control over resources or human activities; the Church and the government became enemies. Local power was then monopolised by the *hacienda* owners, who managed effectively to control the Morelos environments. Since the revolution a new overall domain has been established, again through the implementation of strongly centralised mechanisms of control over the use of resources and the organisation of labour. In discussing the three stages I have attempted to examine both the strategies of those actors involved in the maintenance of the existing domain, and the strategies of actors which arise out of the oppositions implied in the domain. It is by looking at the dialectical interaction between opposing forces that the aspects of the process of domain formation become intelligible, as well as the causes which lead to the disintegration of a domain.

The analysis of overall power does not exclude the analysis of regional domains: the latter directly determine the interdependence of specific human groups and resources. The term 'region' refers to a defined space which presents distinct ecological traits. But it should be remembered that ecology is a human adaptive process: it cannot be analytically divorced from society and history.[16]

Notes

1. *Pueblo*: village, small town.

2. For instance, the research undertaken in the region of Texcoco by students and staff of the Universidad Iberoamericana (Gómez, 1971; Campos, 1973; Pérez Lizaur, 1975; Creel, 1977); the comprehensive analyses of the Tlaxcalan area by Nutini (1965, 1968), Nutini and Murphy (1970) and González Jácome *et al* (1977); the studies of Morelos and Jalisco sponsored by the Centro de Investigaciones Superiores del Instituto Nacional de Antropología e Historia (Warman *et al.*, 1974–6; Arias and Bazán, 1977 and 1979, Fábregas, 1977, Martínez and Gandara, 1976, De la Peña *et al.*, 1977). Outside Mexico one should mention the Central African research organised by the Rhodes-Livingstone Institute, and the Mantaro Valley studies in Peru (Long and Roberts, 1978).

3. A relatively recent monograph on Tonalá (Díaz, 1966), a *pueblo* nowadays engulfed within the urban area of Guadalajara, focuses the analysis on internal social relations, *as if* they existed in a self-contained universe. Guadalajara had over a million people at the time of Díaz's study.

4. My italics. Compare this insular view of the community with the quotation inserted at the beginning of this introduction.

5. See also Wolf (1955 and 1957). The abundance of particularistic anthropological studies on Middle America is testified to by the list of monographs provided by Cline (1952); see also Adam's discussion (1965b: 260 ff.).

6. *Hacienda*: large landholding. *Encomienda*: colonial institution designed to allocate tribute and periodical Indian labour to Spanish entrepreneurs.

7. See, for instance, Kunkel (1965); Dalton (1972: 405). The village chosen by Redfield as an example of 'folk', isolated culture was really 'an armed camp created by the Yucatán War of Castes' in the nineteenth century (Adams, 1965a: 8; cf. also Strickon, 1965). Redfield's Rousseauesque ideas, as well as his view of social change as a result of a process of one-directional diffusion along a folk–urban continuum (1941) have had a persistent influence on the literature, as testified by Evaristo de Moraes Filho's bibliographical article on Latin American sociology of development (1970).

8. For a fuller presentation of modernisation theory, see Smelser (1967). Two critical assessments of this theory are those by Frank (1971) and Long (1977).

9. In Mexico, this would be to ignore the historical process of complex interdependence and feedback brought about by the conquests and merging of tribes in the prehispanic period, by the overall dominance of Spanish rule during the Colony, and by the action of nineteenth-century independent governments and multinational enterprises.

10. However, when he analysed the Russian case in depth, Lenin (1950) pointed out that capitalist penetration into traditional rural areas did in fact use and reinforce non-capitalist forms of organisation.

11. Such *needs* are culturally defined.

12. The word 'segment' was coined by Durkheim; but it has been more recently developed by Julian Steward and his disciples, within a multi-evolutionist theoretical framework.

13. Some examples: Mitchell (1956), Epstein (1958), Bailey (1960), Long (1968), Kapferer (1969 and 1972).

14. The idea of situational selection comes from Evans-Pritchard's *Witchcraft, Oracles and Magic Among the Azande* (1937).

15. 'Extensive' and 'intensive' are my words, but I think they are implicit in Adams' discussion.

16. Two main conceptual models have been used for studying regions in Middle America: Steward's 'levels of integration' (1956) and Aguirre-Beltrán's 'dominical process' (1967). Both models are rather limited: the first one does not analyse interrelationships between 'levels', and the second one stresses inter-ethnic power and its unidirectional effects on rural areas. I found a more congenial source of inspiration in Leach's study of linkages between the Kachin mountains and the Shan valleys (1954), and in Murra's description of 'ecological verticality' in the Andes (1975).

II
THE SETTING

Las geografías también son simbólicas: los espacios físicos se resuelven en arquetipos geométricos que son formas emisoras de símbolos. Llanuras, valles, montañas: los accidentes del terreno se vuelven significativos apenas se insertan en la historia. [Octavio Paz, 1970: 108–9]

Situated immediately south of the Valley of Mexico—and Mexico City—the state of Morelos presents a topography which Oscar Lewis described as 'a sugar bowl, tilted to the southeast, with mountains forming the outer rim and the level, fertile plantations forming the center' (1951: 3). In the northern half of the state the physical distinction between the volcanic mountains and the valley is sharper. People call these two areas the highlands and the lowlands, the cold country (*tierra fría*) and the hot country (*tierra caliente*); and they associate aggressiveness and *machismo* traits with the people living in the hot country (below 1,400 metres above sea level), whereas a peaceful and hard-working character is attributed to the highlanders (Ingham, 1968). In fact, in addition to marked climatic differences, there are also economic and social variations between the villages of the two areas, which relate to their ecology and historical background. Lowland communities were sited on the fringes of, or embodied within, sugar-cane plantations and large *haciendas*. Highland villages were also associated with *haciendas* in the past, but in a different fashion. Even if part of their time was spent working for large landowners, villagers also had an economic basis in small-scale farming. Both areas have for centuries formed a symbiotic unit, in its turn articulated to national and international processes of economic dominance.

At present, the state of Morelos is divided into thirty-two *municipios* (townships), which are grouped into seven *distritos*.[1] There are several highland *municipios* in Morelos; however, the term Morelos Highlands (Altos de Morelos) is more specifically used in reference to a set of villages in the upper north-eastern corner of the state. The region[2] occupied by these villages constituted a single political and administrative unit both in prehispanic times—the jurisdiction of Cuauhtenco—and during the colonial period—the *partido* of Totolapan (Gerhard, 1970: 37). Nowadays the region is formed by four *municipios*: Atlatlahucan, Tlalnepantla, Tlayacapan and Totolapan, which in turn are part of the *distrito* of Yautepec. In order to distinguish it from other highland zones of Morelos, the region is sometimes locally referred to as Altos de Yautepec.

Geography, land use and climate

A distinct geographic zone, the Morelos Highlands extends 332 km^2 between roughly 1,500 and 2,000 m above sea level.[3] Its limits are (*a*) to the north, the high peaks of the Sierra del Ajusco and the Sierra Santo Domingo, which extend along the 19th parallel and separate the state of Morelos from the Valley of Mexico; (*b*) to the west, the Tepoztec mountains, which, following the 99th meridian west, stem from the Ajusco range and separate the region from the Valley of Tepoztlán; (*c*) to the east, the slopes of the Sierra Nevada and a small Ajusco-connected range known as Sierra Tlalnepantla; (*d*) to the south, a sharp slope indicating the beginning of the lowlands. Formed on a basis of igneous rock, the area presents a declivity from north to south, interrupted by small plateaus and eroded hills. It is in these folds of the rocky layer that people have built terraces for cultivation. A terrace consists of a stone wall (*tecorral* or *tecuemites*) which retains alluvium and prevents erosion. Its size may vary from 100 m^2 to half a hectare. During the rainy months a frequent task for the highlanders is to collect the slime accumulated in ravines in order to enrich their terraces. Ever since prehispanic times, cereals and green vegetables have been grown in these pockets of rich, loamy soil, most of which can be cultivated every year without fallowing or crop rotation.[4]

Yet only 46 per cent of the total land in the four *municipios* can be of agricultural use (see De la Peña, 1980: Table 2.1). This land constitutes the most valuable resource of the region. In addition there are extensive pinewoods and *madroñales* (tassel-tree forests) in the mountains of the northern *municipios* of Tlalnepantla and Totolapan, which are also valuable; but only small shrubs grow in the lower hills. As for the pasture land, which covers 10 per cent of the total area, it has never been capable of supporting large herds of cattle, owing to its extreme aridity during six months of the year.[5] Infertile soils may have a variety of uses, from making *adobes* or mud bricks to paving roads. Recently the gravelly soil (*tezontle*) which is found in many hills has begun to be used on a large scale for building purposes. There are also a few small lime pits scattered throughout the area.

The year is sharply divided into a wet and a dry season. The rainy season lasts from May to October, and it brings an average rainfall of 950 to 1,000 mm; but in the months from November to April there is not a drop of rain. The wet months are also the hottest (although the hottest *hours* of the year occur just before the beginning of the rains), with average temperatures of around 30°. Temperatures are favourable for agriculture throughout the whole year; however, the lack of water during the six months of the dry season limits agricultural output drastically (see De la Peña, 1980: Table 2.2).[6]

Topography also contributes to seasonal aridity. The declivity of the area inclines the water to run south, through deep gorges and ravines created by erosion, so that no natural water deposits are formed. In addition, there is not a single river or brook in the four *municipios*. A few springs can be found in the hills, but their yield is exiguous and is in any case wasted for lack of dams and aqueducts. The only existing public well in the whole region is in the town of Tlayacapan, where several basaltic strata had to be perforated in 1970 by heavy modern machinery before sufficient water—only for drinking purposes—could be found at a depth of 230 m (Varela, 1972). Some villages of the *municipios* of Tlayacapan and Totolapan were granted irrigated plots in the neighbouring lowlands by the agrarian reform programme implemented after the revolution; but apart from these fields, which cover only 3.45 per cent of the area, the rest of the cultivable land produces only one yearly crop.

Patterns of settlement

Scarcity of water combined with relative scarcity of land has thus ensured that the crop surplus could never be very high, so that population density would not transcend certain limits.[7] For instance, the total population of the area in 1970 was 17,088, and the average density fifty-one persons per square kilometre. These numbers for the same year were grouped into 3,400 nuclear families; hence the average ratio of land to family was only 3.06 hectares, given that the total cultivable land amounts to 10,404 hectares. On the other hand, ecological conditions permit settlements to be more dense and concentrated than in other rural regions with less intensive agriculture and poorer soil. People are highly concentrated in the four municipal centres, which are towns with two or three thousand people. At present there are also fourteen villages ranging between 100 and 700 inhabitants, and in the southern *municipio* of Atlatlahucan several hamlets scattered along the roads lodge a few families each (Table 2.1). Most of these hamlets have come into being very recently, after the agrarian reform of the 1920s and 1930s. Conversely, some peripheral hamlets which existed in the colonial past have disappeared. All settlements have and have had a fluctuating 'floating' population: merchants, seasonal workers, pilgrims, tourists,

Table 2.1
Towns, villages and hamlets in the Morelos Highlands (1970)

Name of community	*Inhabitants*	*Name of community*	*Inhabitants*
Municipio of Atlatlahucan		*Municipio of Tlayacapan*	
Atlatlahucan	3,590	Tlayacapan	3,588
Tlaltetelco	599	San Andrés Cuauhtempa	657
San Juan Texcalpan	305	San José de los Laureles	609
Tepantongo	206	San Agustín Amatlipac	338
El Cabellito	70	Pantitlán	93
Kilómetro 88	68		
El Zapote	50	Total	5,285
Mochomilpa	41		
Thirteen hamlets with less than 30 people each	178	*Municipio of Totolapan*	
Total	5,107	Totolapan	2,488
		Nepopohualco	789
Municipio of Tlalnepantla		Ahuatlán	214
		San Sebastián (La Cañada)	158
Tlalnepantla	1,823	Tepetlixpita	147
Felipe Neri	409	Nicolás Zapata	135
San Nicolás (El Vigia)	395	El Fuerte	128
Total	2,627	Total	4,059

Source: 1970 national census

fortune seekers. Their number is limited by competition for water, housing and land.

In common with other communities of central Mexico, every town or village has a central square (*plaza*) and a central church or chapel. In the towns the central church is adjacent to a central monastery, which in colonial times belonged to the Order of St Augustine. (Nowadays these monasteries or convents are controlled by secular priests of the Cuernavaca diocese.) The streets of all settlements—except the small hamlets— are arranged on the so-called Spanish grid model, although in some places (such as the town of Tlalnepantla and the villages of San José de los Laureles and Nepopoalco) the irregular topography reduces this symmetry.[8] Bay trees, eucalyptus and *amates* (a kind of fig tree) grow on street corners and squares. The principal towns are divided into four homologous sections or *campan*, as in the old Aztec urban form (Lombardo de Ruiz, 1973:52 ff.), by two main streets which intersect at the central square and run from north to south and from east to west. In turn, each *campan* comprehends several *barrios*: clusters of households with a distinct chapel or shrine.[9]

Settlements can be physically classified according to their size—towns and villages—and also in terms of two types of urban composition and housing design. In the first type, houses are built contiguously, so that, with a few exceptions, land within a settlement is not used for intensive cultivation. Most dwellings are made of mud brick (*adobe*) walls with a tiled roof; they have a large rectangular front room, a kitchen and a back yard. The front room opens directly on to the street, and is used for all purposes, though there are a few larger houses which have extra bedrooms. The back yard is rather small; it usually contains a few fruit trees (a Mexican variety of plums, limes, sweet lemons, occasionally oranges, coffee shrubs, guavas and pomegranates), as well as flower beds, poultry and one or two pigs, all kept for domestic consumption or for small-scale sale in local markets. The yard is also used to keep draught animals and store agricultural tools. *Coscomates*—egg-shaped granaries, built on mud—are frequently found in the yard. Towns and villages in the *municipios* of Atlatlahucan and Tlayacapan tend to conform to this first type.

In the second, a good deal of the land within the urban perimeter is used for cultivation; above all, for the growing of peaches and avocados for commercial purposes. The town of Tlalnepantla is the best example of this: houses are more or less compact around the central square; but, as streets leave the centre, terraces with large gardens and orchards dominate the urban landscape—the presence of terraces thus making the transition from agriculture to residential use less abrupt. Dwellings are inside the orchard and do not open directly on the streets, although they are built of the same materials and their internal arrangement is not very different from those described above. The back yard does not exist as such; its functions are served by the orchard, which also contains poultry and pigs, and is used for storing tools and keeping draught animals. Animal manure—together with human ordure—is used as a fertiliser. The villages of Felipe Neri and San Nicolás can also be considered within this second type. As for Totolapan and its satellites, they have intermediate characteristics.

In all types of settlements, most people's houses lack urban-type modern facilities.[10]

Provision of drinking water

Many towns and villages are sited in the vicinity of hills, where *jagueyes* can be built more easily. *Jagüeyes* are the traditional deposits of drinking water. Pits are dug in the ground and channels constructed which bring rainwater down from the hills. Some *jagüeyes* are more than 10 m deep and have a diameter of 25 m. Every person living in a community is entitled to use this communal water. Members of the community are also obliged to co-operate in the cleansing and repair of the *jagüey* and its channels. This task is carried out just before the rainy season.[11] Many houses also have small cisterns which collect rainwater by means of roof channels. The oldest ones are made of stone; nowadays they are built of brick and cement. During the dry season, people who do not want to use the *jagüey* buy clean water which is brought from wells and springs in the lowlands—if they can afford it. Muleteers were in charge of this service until they were displaced by lorries from the near-by towns of Cuautla and Yautepec. Drinking water is also stored in barrels and big clay vessels.

In San José de los Laureles and Tlalnepantla a prehispanic system for collecting water from mountain filtrations is still used. It consists of excavating channels (called *canoas*) in a steep rock strategically situated in a dell, thus leading the water, drop by drop, into a vessel. These *canoas* are usually outside the town; this fact, added to the small volume of water collected, indicates that they were never a real answer to water scarcity.

Aqueducts to bring water from mountain springs would have been a better solution but as far as I know there was only one aqueduct in the whole region, and it is no longer in use.[12] It carried water from a source (or *ojo de agua*) near the town of Tlayacapan, on the slopes of the Tepozteco sierra. A fairly big stone cistern—a *caja de agua*, 'water box', identical to the stone cisterns found in many parts of central Mexico—capable of holding perhaps twenty cubic metres, collected water from the source. The aqueduct stemmed from the cistern and, supported by *tecorrales* and stone arches, took water into the centre of the town. Tradition maintains that this aqueduct was there before the Spaniards came, and that it was later kept in good shape by the friars. The water was used for drinking and washing; there was not enough for irrigation on any significant scale. Towards the end of the last century the source began to diminish. After the revolution it appeared to be almost exhausted. People thought that stones and mud might be obstructing the flow of water, and dynamited the hole. The result was disastrous: the source dried up altogether. An old woman said that 'the lords of the mountain' had spoken to her in dreams: they wanted a young Tlayacapan woman to sacrifice her newly born son with a knife over the cistern; water would start to flow again as soon as the blood touched the stone. To date, no one has volunteered to perform the ceremony.[13]

Interdependence of mountains and valleys

The Morelos Highlands have never been an isolated region. Numerous roads and footpaths have for centuries allowed communication with all neighbouring places. It was the boast of the muleteers to cross the Ajusco sierra in a single day's journey. Carrying all types of commodities, they travelled fifteen hours among forests and precipices before arriving in Xochimilco—which was already within easy reach of Mexico City. Nowadays the trip takes about half an hour, thanks to a modern highway and the motor car. Western paths had a mystical meaning: Quetzalcoatl-Tepoztecatl, the god king who discovered *pulque*, was born in those mountains.[14] His shrine still stands on the outskirts of Tepoztlan, even though the roads passing by have been 'christened': people go along them to visit the shrine of the Lord of Chalma.[15] Conversely, pilgrims coming from the west take the same route to visit the miraculous images of Los Altos: the black Christ of Exaltacion and Our Lady of Tránsito in Tlayacapan; the Señor Aparecido ('the Lord who turned up') in Totolapan; St Joseph in San

José de los Laureles.[16] In addition to this religious dimension, the same western paths have always been used as trade routes to Cuernavaca—the capital of Morelos state in modern times—and to the state of Mexico. As for the eastern paths, they also led—and lead—to places of commerce and devotion: the towns of Ozumba, Amecameca and Tepalcingo, which have both important markets and famous shrines. But most important, as in the past, are the trails and roads to the south. In a sense it is because of this intense intercommunication that certain crucial ecological differences between highlands and lowlands have been reinforced—if not entirely created.

This point will emerge more clearly if we look at the most obvious contrasts between the two areas. I will draw two 'ideal models', of highland and lowland economies, in order to highlight these contrasts. In the Highlands, agricultural soil offers itself, as it were, in small quantities. The intense, small-scale technology of Highland terraces is essential to avoid erosion or even total loss of land. Sometimes heavy rains destroy a good part of the terrace soil; people then have to buy soil—or steal it— from a neighbour, or from a deserted field, and bring it to the devastated plot by truck or donkey. The household appears to be an ideal unit of labour and exchange of know-how. Human settlements are essentially an aggregate of families who farm their terraces. Neighbours know and watch one another. Strangers have no place in such a community: there is not much to do, apart from cultivating maize and pulses on the family terrace. This is of course a stereotype, inadequate to describe the complexity of social life in the biggest Highland towns; it is, however, fairly applicable to some of the smaller villages and useful as a point of comparison with the lowlands.

The size, the climate and the humidity of the hot-country valleys seem to encourage large-scale, plantation-type production. And, in fact, this type of production has existed for over four centuries. Labour is needed in massive contingents. Therefore its organisation cannot be contained by the household structure. Human settlements are open ended: everybody is admitted to work in the harvest. Neighbours are very often strangers to each other, coming today and disappearing tomorrow. Agriculture is not the only possible occupation. Most people do not grow their own food: commerce and services are sources of work almost as important as agrarian activities. Hot-country towns and villages do not present the compact, symmetrical patterns which are common in the Highlands. Yautepec and Cuautla, for instance, two sizable towns which dominate the commerce of the whole north-eastern part of Morelos, have the aspect of enormous open markets—dirty streets, vendors everywhere, hundreds of buyers coming and going, constant noise. Both towns have always outnumbered the population of the largest Highland towns.[17]

Another striking difference between highlands and lowlands can be observed in the use of water, and in the general attitude that people have towards it. The valley benefits from the complex system of the Amacuzac river, as well as from numerous springs and relatively shallow wells. Underground currents are partly fed by the Highland rain torrents. Irrigation is easily available. And water is also used for relaxation and amusement. Oaxtepec, Cocoyoc and Cuautla have for centuries been tourist resorts—the Aztec rulers themselves used the valley of Morelos for this purpose.

Yet it is important to notice that it is precisely the lack of water in the Highlands which has permitted the flow of a fair number of seasonal workers to the lowlands.[18] It is surprising that, except for small-scale systems which convey drinking water, hydraulic works are absent in the Highlands—whereas traces of such works are found everywhere in Mexico; furthermore, they were a crucial characteristic of prehispanic and colonial economic and political organisation (Palerm, 1972: Palerm and Wolf, 1972). An amazing example of Aztec hydraulic technology was the diversion of the Amecameca rivers from their natural beds towards Morelos (Palerm, 1973: 187ff.). The main branch of these diverted rivers benefited some land in the town of Atlatlahucan (Rojas, 1974: 67 ff.). In the early seventeenth century the Spaniards returned the river to its original course, which ended in Chalco lake; but oral tradition in Atlatlahucan keeps a living memory of the river and its effects on local agriculture. Apart from this technological feat, there seems to have been no attempt to irrigate the area, in spite of the feasibility of constructing dams in the ravines and 'water boxes' in the vicinity of mountain sources. Such small-scale irrigation systems can be found in the eastern towns of Zacualpan and Hueyapan, which are situated on the slopes of Popocatepetl and have a topography comparable to the Highlands (López, 1974: 31 ff.; Neyra, 1976).

People in the Highlands believe that this lack of irrigation is the result of a deliberate design. They say that without the rain torrents escaping from the mountains towards the south the wealth of the Yautepec river and the Oaxtepec springs might be diminished. To confirm this hypothesis, local people draw the attention of the visitor to what is for them a crucial fact: the streets of towns and villages were designed in such a way as to lead rainwater not to local cisterns but to canyons leading down into the south. History is then conceived as an on-going conspiracy aimed at depriving them of a valuable resource. The high value attached to this resource is obvious in the legend of the Tlayacapan spring.

In reality, rainfall originating in the Morelos mountains feeds not only the Yautepec river but also the Nexpa, the Tenango, the Chinameca (Cuautla), the Xochitepec (Cuernavaca) and the Tembembe. These rivers have been, and are, used in the irrigation of sugar-cane fields (Hernández-Orive, 1973: 20). Conspiracy or not, a plausible hypothesis is that there is a pattern by which the water has been concentrated in the lowlands so that the valley should be able to specialise in large-scale plantations and irrigated crops, whereas the uplands should provide grains and a seasonal labour force. This pattern is a function not of nature but of a political decision, renewed throughout different periods of history.

Notes

1. A *municipio* is a territorial administrative unit comparable to the British rural district; it includes a head (*cabecera*) or principal town, bearing the same name as the *municipio*, and several subordinate villages. A *distrito* is a compound of several *municipios* with a common court and a common congressional deputy.

2. *Region*: 'Tract of country, space, place, of more or less marked boundaries or characteristics' (*The Concise Oxford Dictionary*.) I will use the term interchangeably with the term 'area'.

3. For the British equivalent of this and other units of measurement used throughout the book, see Appendix II, Table 1.

4. Local agriculturalists distinguish between three types of alluvial soil: (*a*) *atocles* or *tierra amarilla* (yellow ground), which is rather hard, with few interstices and a low degree of water retention; (*b*) *tierra negra* (black ground), which is softer and less dry; and (*c*) *tierra arenosa* (sandy land). The first type is more suitable for modern cultivation because it retains chemical fertiliser better; the second type is easier for tillage, but fertiliser tends to permeate through it, and is occasionally lost. The third type is looser, less productive, and may need rotation every two or three seasons. (Rotation, however, is deemed to be convenient for all types of soil after being used for growing tomatoes with modern techniques). Most agricultural plots have soil of the first and second types. Higher terrains are more stony and eroded; lower terrains tend to be of the second type. Infertile lime soils are called *tepetates*.

5. The amount of pasture land has decreased throughout the years, due to excessive grazing and erosion. Forests have also suffered excessive cutting. This has resulted in the expansion of barren soils.

6. As the minimum temperatures only occur a few nights in the winter, frost tends to be rare—especially in the *municipios* of Atlatlahucan and Tlayacapan; conditions

are less favourable in Tlalnepantla and Totolapan, which are exposed to northern winds and have a slightly colder climate. In any case, frost comes *after* the harvest. The amount of rainfall is enough for good seasonal agriculture; but not enough to produce floods. The average rainfall is 717 mm; the maximum is 2,023 (in Tabasco); the minimum, 340 (in Sonora). See Tamayo 1962, vol. II: 160.

7. For a discussion on the relationship between irrigated and non-irrigated agriculture and size of rural settlements, see Palerm (1952), Sanders (1967), and Pérez-Lizaur (1975).

8. The grid model was probably prehispanic; also the present outline of most Highland settlements. Favier (1977) has shown that space proportions within the urban perimeter of Tlayacapan are identical to space proportions in Aztec Tenochtitlan.

9. The town of Totolapan does not have any standing *barrio* chapels. Yet the division into *barrios* does persist, at least as a way of referring to different sections. It should be emphasised that territorial divisions are social and political devices which vary historically.

10. According to the 1970 census, only 5 per cent of the houses in the four *municipios* had a proper sewerage system, and less than 10 per cent had toilet or latrine. Although electricity could be found in 10 per cent of the houses, electric appliances were rare, with the exception of radio and (to a lesser extent) television sets. Water pipes only existed in the town of Tlayacapan.

11. The tradition of communal work seems to be disappearing in many places. More and more people prefer to get water by individual means. In San José and San Agustín *jagüeyes* have been left to deteriorate so that they are practically useless.

12. My friend Claudio Favier, an architect who has done urban research in Morelos, drew my attention to the existence of this ruined aqueduct. Later I tried to find others in the Highlands, but unsuccessfully.

13. This is reminiscent of cults of Tlaloc, the Aztec rain god.

14. To the traveller's surprise, a big poster in Amatlan, a little village lost in the Tepozteco sierra, announces: 'Ce Acatl Topiltzin Quetzalcoatl in this place walked his first steps with his gold sandals'.

15. Chalma is a nationally famous shrine lodging an old sculpture of Jesus on the cross (Turner, 1974: 190 ff.). The pilgrimage route in Morelos goes through Tlayacapan, San José de los Laureles, Amatlan, Tepoztlan and Huitzilac.

16. The Lent *fiesta* of Our Lady of Transito in Tlayacapan is organised and paid for by people from Tepoztlan. Lent pilgrims bring money to the Christ of Exaltacion. Similarly, a group comes every year from San Andres Ahuayucan (state of Mexico) to San José de los Laureles and provides St Joseph's image with presents: flowers, candles and money. These and other religious festivals will be analysed in Chapter VII.

17. In 1970, Yautepec had 26,000 inhabitants, and Cuautla 68,000.

18. The valley of Morelos has always been a pole of attraction for migrants from everywhere in Mexico. Yet the most constant supply has always come from the Highlands. This is especially the case for the region Cuautla–Oaxtepec–Cocoyoc–Yautepec.

III
AN APPROACH TO REGIONAL HISTORY

In this chapter historical information is presented to demonstrate that the Morelos Highlands emerged as a region by virtue of their symbiotic relationship with the neighbouring lowlands. It is also contended that this symbiosis has had different characteristics and effects in different periods. These periods are in turn defined and measured in relation to the structure of the larger society. In the colonial period the centralised power domain conditioned a type of regional interdependence such that persistent conflicts did not seriously threaten the established order. However, national independence entailed the fragmentation of the overall domain and the emergence of conflicting regional interests which culminated in the Zapata revolution. (In later chapters we shall see how post-revolutionary governments have attempted both to re-establish a centralised domain and to solve the conflicts).

In the light of regional history, a critique is made of an analytical tool often used by Central American anthropologists: the model of the corporate community. This critique lays the foundations for a discussion of local conflicts and changes. There is also a section on colonial *fiestas* and their transformations during the nineteenth century. *Fiestas* have often been considered as symbols of communal cohesion and as mechanisms for equilibrium; but in this book they are seen in the context of both regional and national oppositions and heterogeneity. In this and other chapters, the analysis of ceremonies and rituals is used to highlight the basic components of social processes, in so far as the use of religious symbols by different actors and social categories provides a crucial index for understanding their interests and strategies *vis–à–vis* one another.

The emergence of a colonial society

Prehispanic background. The Nahuatl names of settlements, hills and ravines dispersed over the territory of the Highlands bear witness to ancient culture.[1] However, the oldest language in the region was not Nahuatl but the tongue spoken by those roving hunters who came to Morelos perhaps eight thousand years ago. About these people we know little more than that they existed.[2] The supremacy of Nahuatl-speaking groups in Morelos coincides with the flowering of Tolteca civilisation, in Tula, towards the ninth century A.D. Together with a language which was to become a sort of *lingua franca* in the whole of central Mexico, this civilisation brought agriculture, textile crafts, pottery, urban architecture and what we would call 'economic planning'. The extraordinary calendar introduced by the Toltecas was a precise charter for agricultural activities (Florescano, 1971: 21); it was also an example of the geometrical knowledge which created monumental cities and complex systems of irrigation and transport. The utopian new society that the Spaniards would bring to life in the sixteenth century had a solid foundation in the Nahuatl culture. Many of its elements have persisted until this day, combined with new elements in different structures.

Two Nahuatl groups settled in Morelos: the Tlahuicas, who controlled the hot country, and the Xochimilcas. The territory of the latter covered the whole Ajusco range, from Lake Xochimilco—at the southern extremity of the valley of Mexico—to the Tepozteco sierra and the southern Ajusco slopes. Their distribution upon both sides of the range allowed them to become a crucial link in the commercial exchange between the central valleys and the distant south—the regions of Oaxaca, Guerrero, Veracruz, Puebla (Gerhard, 1970). The principal Xochimilca town in the Morelos Highlands was probably Tlalnepantla-Cuauhtenco; but it is likely that most settlements in the region were also founded by the Xochimilcas.

In the fifteenth century the Mexicas—another Nahuatl-speaking group, which was to create a military confederacy: the so-called Aztec empire—incorporated Morelos into their domain (Barlow, 1949: 80; Gibson, 1967: 17). Some places in the Highlands—Tlayacapan, for instance—became strategic military points for the whole empire: the peaks and hills of the area are natural watchtowers dominating the view of the southern valleys. The ruined stone fortresses which can be seen in some mountainous sites were probably built by the Aztecs.[3] On the other hand, the new military rulers also had an interest in agricultural planning: they diverted the Amecameca rivers into Morelos and channelled labour from the mountains towards the hot-country valley. A strong interdependence of lowland and highland communities thus emerged, and resulted in the development of cotton and cocoa plantations in the irrigated valleys—particularly in the zone of Cuautla-Amilpas, the produce of which was taken direct to Tenochtitlan, the Aztec capital city (Carrasco, 1976b: 183). Oaxtepec, a lowland town with generous springs, was chosen by the Aztec nobility for establishing both recreation resorts and centres of local political control. Oaxtepec was the point where the Highland villagers had to take their tribute for the emperor: cotton blankets, turkeys, maize and beans, feather shields, paper made of amate-tree pulp—and seasonal labour.[4]

Conquest, marquisate and ecomiendas. Fernando Cortés arrived on the Mexican coast in April 1519 and reached Morelos two years later (Riley, 1973: 10 ff.). He sent an expedition in January 1521, which took Yecapixtla and Oaxtepec—where the conquerors found orchards and gardens 'such as they had never seen in Europe' (Díaz del Castillo, 1974: 306). Soon after, Cortés personally led an army which was to subdue Morelos altogether. It was clear that Tenochtitlan would be more vulnerable if products from the south stopped reaching the city through the Ajusco sierra (Martínez-Marín, 1968: 26).

In the Highlands the Xochimilcas, perched on hills and steep rocks, attacked the conquerors with thousands of arrows and stones. Cortés and his men suffered acutely from thirst and could find no water anywhere (Cortés, 1975: 122 ff.). A large number of Xochimilcas were concentrated in Tlayacapan: there Cortés decided to assault the fortfied cliffs. One of these was sited above a spring which provided the town's water—probably the same spring described in Chapter II. Towards sunset the Spaniards, tired and thirsty, succeeded in capturing the source and stayed there the whole night, enduring the missiles and insults thrown at them from the surrounding hills. In the morning Cortés opened fire; several hours later, harquebuses and thirst forced the Xochimilcas to surrender. After the fall of Tlayacapan—the main focus of resistance in the region—other Highland settlements also gave up the struggle (Díaz del Castillo, 1974: 312 ff.). After this victory Cortés went down to Oaxtepec and was lodged in a local overlord's house, where he was again struck by the beautiful gardens, the variety of flowers and fruits, and the fertility of the fields (Cortés, 1975: 124).

Quauhnahuac, the main city of the valley, which the Spanish tongue was to transform into Cuernavaca, fell soon after. Four months later, Tenochtitlan was defeated for good. History made a decisive turn. Aztecs, Xochimilcas, Tlahuicas, all other American dwellers and their descendants would henceforth have a new name: Indians. As for Fernando Cortés, he did not forget the wealth and charm of the southern valleys he described so vividly in his fifth letter to Emperor Carlos V. He decided that Morelos would be a place for his personal enjoyment and benefit. As early as 1522 sugar cane was planted, on his instructions, in the vicinity of Cuernavaca, and so Morelos's lot was drawn for the coming centuries.

Cortés's dominion over Morelos was soon a *fait accompli.* The land he acquired through royal *mercedes*[5] was entailed to his new title: Marquis of the Valley of Oaxaca. Morelos was—qualitatively if not quantitatively—perhaps the best part of the marquisate, and in Cuernavaca he built a magnificent palace. He sought glory and honour to rival that of the ancient Castilian nobility; yet his mind was nearer a modern entrepreneur's than a feudal overlord's. In his realm Cortés promoted the expansion of traditional crops such as cereal agriculture and cotton. More important, he also created a great many new economic activities: sugar-cane plantations, livestock, fruit trees, wheat production and flour mills, European-style textile workshops (*obrajes*) (Riley, 1973: 51 ff., 66 ff.). For all these enterprises a labour force was provided in three ways: through the occupation of Indians working full-time; through the purchase of Negro slaves brought from Africa via the Caribbean Islands, and, more frequently, through *encomienda* Indians.[6]

Some authors (e.g. Semo, 1973) have seen this institution as the result of European feudalism transplanted to America. However, *encomiendas* increased the efficiency of Morelos enterprises to participate in a competitive market and led to a genuine reproduction of capital. It should be remembered that the colonial economy in Mexico had two important sectors. One was entirely geared to export to Spain: the extraction of precious metals and the production of specialised items; the other was devoted to produce for the internal market: to feed and clothe workers in mines and plantations, as well as bureaucrats in cities. This second sector was forcefully stimulated by severe restrictions enacted by the Crown on imports from overseas.[7] Morelos had no significant mining activity, but produced sugar both for export and for internal markets, and a variety of items for the Mexican towns and cities.[8]

Encomiendas were functional for enterprise in the sense that they secured seasonal labour, which did not involve the expenses of full-time workers or even slaves—the latter had to be fed the whole year round, whereas *encomienda* Indians were seasonally sent back to their communities, where they grew their own food. It was important to save money for the heavy investments required in sugar-cane fields: irrigation, roads and levelling.

Up to 1532 Cortés managed to get *encomienda* Indians from practically all communities in Morelos; after that year New Spain's second Audiencia (Supreme Council) decreed that the villages of the Morelos Highlands should be separated from the marquisate in order to become the *partido* (district) of Totolapan (Riley, 1973:33).[9] Yet most cold-country settlements continued to provide labour for other *encomenderos* of the valley who had set up enterprises comparable to those of Cortés.[10] Later, from the seventeenth century onwards, Highland labour was also channelled into the southern lowlands by means of the *repartimiento* (literally, 'distribution'), an institution analogous to *encomienda,* geared both to the execution of public works and to private enterprises. An important *repartimiento* centre for Highland Indians was the hospital of Oaxtepec, which was not only a charitable institution but also the owner of sugar-cane plantations and sugar mills (Zavala and Castelo, 1939–45, vol. V, Nos. XXX and CLI). What is important to stress is that both *ecomienda* and *repartimiento* reinforced the Aztec-designed interdependence, combining resources from valley and mountains for the benefit of alien overlords.

Colonial religion, political organisation, and economics. The 'spiritual conquest' of Los Altos was carried out by friars of St Augustine's Order after 1530.[11] The first Highland monastery was founded in Totolapan—the main town of the *partido*—in 1533 or 1534. Its prior, Fray Antonio de Roa, is said to have performed miracles there in 1543, thus converting the whole population to Christianity (López-Beltrán, 1969). The monasteries subsequently founded in Atlatlahucan (1554) and Tlayacapan (1560) are magnificent witnesses to Augustinian local power and prosperity (Ricard, 1947: 99, 195).[12]

The power of the Cross reinforced that of the sword and of the new law. Even before the friars came to the Highlands, there were Spanish soldiers and royal bureaucrats. The latter, as well as other Spaniards who had acquired property in the region[13] and settled locally, used the energy and dexterity of indigenous workers to build both government quarters and private houses. Many of these impressive buildings have resisted the erosion of centuries and stand in the vicinity of central squares in Highland settlements.

It is interesting that the size of private land holdings in the Highlands was never significantly large[14] (with one exception, the livestock *hacienda* San Diego Huixtla, which occupied the land belonging to the town of Totolapan). In contrast, estates in the southern valleys were large from the very beginning of the colonial period; in addition, their exaggerated expansion engulfed Indian communal land (Sotelo-Inclán, 1970: 36–7; Hernández-Orive, 1973: 16, 57 ff.; Warman, 1976: 43). This fact seems to support the hypothesis that there existed a systematic division of labour between hot country and Highlands. Land—especially agricultural land—in the latter was kept available for the Indians, so that they could grow their own food and provide a surplus of grain to feed permanent plantation workers. In their turn, Highland Spanish landowners specialised in raising cattle, destined for consumption in the lowlands. Forests were also systematically cut down by the Spaniards, to be used as fuel in sugar mills.

In 1579 Andrés de Curiel, *corregidor* (magistrate) of Totolapan, wrote a detailed report on his district, listing Totolapan, Tlayacapan and Tlalnepantla as *cabeceras* (heads) or principal towns (Curiel, 1905).[15] He also mentioned smaller villages, or *barrios*, and *congregaciones*: new settlements which resulted from congregating hamlets and dispersed homesteads for the purpose of increasing control over the population. All human settlements, old and new, were organised by the colonial authorities into the regime called *republica de indios*, which allocated houses and communal land to Indians, and allowed royal officers to keep a precise registry of people and production.[16] The structural consequences of this regime will be discussed below. But at this point it is important to stress the interdependence of land tenure and housing distribution, on the one hand, and ethnic definitions, on the other. By and large, Indians (*naturales*), were those who were registered as members of a community and had the right to use a plot in the communal property. Non–Indians (*gente de razón*, reasoning people) were those who could occupy bureaucratic posts, set up enterprises, and hold land as private property.[17]

Social differentiation. From the end of the sixteenth century onwards

Tlayacapan grew and developed faster than other places in the Highlands. In 1570 the townspeople asked for autonomy from Totolapan—a request that was granted a few years later (Riley, 1973: 116, n. 22). The town was given the title of *villa* by the Viceroy at the beginning of the seventeenth century, and later became the head of a large district which included Totolapan itself.[18] The political prosperity of Tlayacapan was associated with the emergence of economic activities which developed side by side with food production and labour provision for the lowlands. These activities produced a surplus, but did not require the use of land or large quantities of labour—and consequently they did not threaten the symbiotic pattern. Commerce, for instance, was an important occupation. The old road which crossed the Ajusco was kept in use during the colonial period, even if passengers and large-scale cargo going to the far south now used the eastern Chalco-Amecameca route. (The wheel, that striking innovation brought by the Spaniards, demanded less abrupt roads. The town of Atlatlahucan, which is the nearest to the eastern road, was one of the stops in an eighteenth-century stage-coach and cargo route between Oaxaca and Mexico City. See Tamayo, 1962, Vol II : 60). Muleteers came and went from Mexico City every day, carrying a variety of commodities. They also wove a network of exchange among the Highland communities and the south. Tlayacapan itself provided two important local products in addition to agriculture: wax candles and ceramics. Wax was collected from beehives dispersed over the hills and fields. It was used to make candles for domestic and Church use, as well as for manufacturing *escamadas*, colourful, baroque flowers made of wax—a traditional decoration for chapels in local religious festivities. In the eighteenth century the wax craft became an industry: production was carried out by several enterprises occupying full-time labourers. One of these was lodged in an enormous stone house next to the *plaza*, still known as 'the wax factory' (*la cerería*). As for ceramics, the fame of Tlayacapan potters was paramount in Morelos and the region of Chalco-Amecameca. They lived in the eastern *campan* (quarter) of the town (called Texcalpan). In the early seventeenth century they built the chapel of Santiago (St James), the most beautiful of the numerous local chapels. In addition to the ordinary market, for which cups, dishes and vessels were provided, Tlayacapan ceramics met a ritual demand: that of incensories, candlesticks and 'pottery for the dead'.[19] Pottery was mainly an occupation of Indians. In contrast, wax craftsmanship and commerce included Spaniards and *mestizos* (half-castes).[20] These three occupations provided channels for individual enrichment and therefore led to internal stratification.

Another source of individual ascent was small-scale mining. Several drifts were opened at the bottom of a local canyon, following a copper vein—which, apparently, was not very generous. Other small veins of copper, silver and gold were discovered and exploited in the surrounding hills. Local versions of history maintain that the miners were both town dwellers and foreigners, and that the large hostel behind the monastery was often full of them—and also of roving merchants.

In any case, what is patent is the social heterogeneity existing in Tlayacapan. Other places in the Highlands also presented internal social differentiation—though to a lesser degree—based on activities which had the possibility of producing surplus for individual wealth. Tlalnepantla, for instance, provided fruit for the market, and Totolapan roof tiles. All settlements had craftsmen: butchers, tanners, saddlers and cobblers, who worked the hides of the cattle raised by Spanish landowners; bakers, masons, even one or two silversmiths, working on silver collected locally.[21] All these elements of differentiation should be taken into account for the understanding of social organisation in the colonial Highland society.[22]

But, again, I want to stress that differentiation occurred in the framework of the symbiotic unity between mountains and valleys, and that it is this framework that created a distinctive social pattern in the region. The pattern persisted during the whole colonial period. The basis for this contention is the continuing success of the sugar-cane business (see Table 3.1). Except for a rather brief period of crisis in the late sixteenth and early seventeenth centuries, the market for sugar expanded, slowly but surely, both at the national and at the international level. This permitted not only the expansion of *haciendas* and the concentration of capital in the lowlands, but also the consolidation of the

Table 3.1 Sugar and population in Morelos during the colonial period

	1600	*1800*
Number of mills (and/or refineries)	13	36
Hectares of sugar cane (cut)	240	3,685
Morelos population	55,000	61,000

Source: Barrett and Schwartz, 1975: 552

sugar business as a pole of social and economic organisation in the whole of Morelos (Hermández-Orive, 1923). Between 1600 and 1800 the number of mills tripled, and the amount of land devoted to sugar-cane cultivation increased fifteen-fold (Barrett and Schwarz, 1975:552). This means that the mills underwent a process of considerable technological sophistication. On the other hand, the population hardly increased at all, in large part because of the ravages of smallpox and other diseases imported from Europe.

The model of the corporate community: a critique

In recent Central American literature the paradigm or model most consistently developed for the sociological analysis of compact, long-established rural settlements living on intensive agriculture[23] is that which views communities as corporate social entities constituting 'veritable redoubts of cultural homeostasis' (Wolf, 1965: 88).

> The distinctive characteristic of the corporate peasant community is that it represents a bounded social system with clear-cut limits, in relation to both outsiders and insiders. It has structural identity over time. Seen from the outside, the community as a whole carries on a series of activities and upholds certain 'collective representations'. Seen from within, it defines the rights and duties of its members and prescribes large segments of their behaviour. [Wolf, 1955: 456]

The basis of the corporation—according to the model—is the fact that members of the community are co-owners of a tract of communal land. Given this basis, the allocation of rights and duties to individuals depends on the existence of a normative consensus supported by a number of features balanced in stable equilibrum.[24] Paramount amongst these features are communal jurisdiction over the disposal of land and the utilisation of the land chiefly for consumption purposes. Other features have the function of preventing disequilibriums. This accounts for both the attitudes of defensiveness and distrust towards the introduction of innovations, and the politico-religious structure existing in the locality. By this structure, the members of the corporation (i.e. the co-owners of communal land: the Indians) who occupy positions of local leadership and authority are obliged to organise and finance festivities (*fiestas*) in honour of the patron saints of the community. In a situation of general poverty, where economic surpluses are available to only a few individuals, the *fiestas* are an instrument of wealth distribution which prevents economic differentiation while simultaneously reinforcing the traditional system of prestige ranking and validating political authority.

The relationships among community members are thus conceived of as being embedded in a system of reciprocal exchange, whereby individuals who are wealthy by local standards distribute their wealth, receiving in return political offices (*cargos*).[25] At the same time, 'the political religious system as a whole tends to define the boundaries of the community and acts as a rallying point and a symbol of collective unity' (Wolf, 1955: 457). Thus, by providing both a culturally approved means of enhancing status and a routinised way of stressing obedience to communal norms, the *fiesta* organisation is regarded as the most important institution for 'the maintenance of an already established sociocultural order' (Cámara, 1952: 146), or, using Wolf's words again, for the maintenance of the community's 'structural identity over time'.

But for how long is this *structural identity* assumed to have existed? The features of the corporate community are considered by some writers to be a function of the survival of prehispanic culture and social organisation (Aguirre-Beltrán, 1953: 21 ff., 39 ff.; Aguirre-Beltrán and Pozas, 1954: 186–7). Others (Wolf, 1957, 1965; Carrasco, 1961), however, have convincingly argued that, even if communities contained many cultural traits that can be considered 'Indian' or prehispanic, the corporation *qua* social structure was a creation of the Spanish government and served a manifold colonial purpose. From the point of view of the colonial authorities, these communities provided a means of controlling the natives and collecting taxes from them easily, as well as of facilitating mass evangelisation. On the other hand, the demographic catastrophe suffered by the Indian population at the end of the seventeenth century through maltreatment and diseases brought from Spain made the colonial government aware of the need to prevent excessive hardship among labourers. Labour from the communities was used by Spanish enterpreneurs without the Crown losing control over it, through the already mentioned institutions of *encomienda* and *repartimiento*. Indian labourers were not supposed to lose residence in the community, and the whole operation of recruiting Indians and mobilising tribute was mediated through royal officers. The law prescribed residential separation between Indians and non-Indians. Indians had to be registered in a locality, and could not move around from place to place (Zavala and Miranda, 1954: 38–9). Thus an essential part of Crown policy was that community boundaries were intended to remain 'untouched', and links with the outside carefully regulated and restricted.[26]

One should question the validity of using an administrative

project—the *república de indios*—as a scientific model for social behaviour, especially since the project failed in its main objects. Consider only one instance: the principle of using communal land exclusively for consumption—which also means that consumption needs were basically satisfied by local production, and not by exchange. This was based on a false assumption: that local production in communal land was always enough to feed the villagers. We know that agriculture in New Spain suffered gravely from cycles of poor harvests and famines, and that in critical years the affected Indians had to become 'desperate buyers of grain' (Florescano, 1971: 126). Even in years which were not so critical, villages without access to irrigated land, which could not produce the whole year round—like those in the Morelos Highlands—sometimes suffered from scarcity of cereal, the basic foodstuff. On the other hand, this scarcity, in the wider society, was often artificially created and fostered by the big *hacendados*' monopolising manoeuvres (Florescano, 1969: passim; 1971, 120–1, 211). These economic crises made it impossible in practice to maintain strict community boundaries and to regulate the movements of Indians, as Indians had to look for alternative ways of earning their living. This fact was surely related to the abundance of permanent or semi-permanent Indian labour in *haciendas*, textile workshops, mines and servile domestic activities (Gibson, 1967: ch. 9). In their turn, at times of economic expansion, Spanish entrepreneurs were anxious to recruit workers without bureaucratic limit—and there were many ways of getting away with it, such as bribing royal officers (Sotelo-Inclán, 1970: 36, 98ff.). At times of hardship Indian labourers were sent back to the villages, or to increase the number of the always abundant urban lumpenproletariat (Gibson, 1967: 392). On the other hand, the products of the village land were often demanded by entrepreneurs to feed their full-time labourers, or else the actual use of the land was controlled by entrepreneurs to grow cash crops. It is reasonable to assume that these pressures had important repercussions in community social life. All this makes the idea of the structural continuity of village life debatable, to say the least. In any case, continuity has to be found in the wider mechanisms which kept the colonial system together.

In other words, it is of paramount importance to stress that the structure of village life was contingent upon a set of changing circumstances in the wider society. These circumstances ranged from climatic and demographic conditions to variations in the need for labour caused by fluctuations in the demand for European and national markets; from the landowners' ability to monopolise grain to the morality of Spanish authorities. Sometimes the combination of circumstances favoured the trend towards a higher degree of local stability; but other periods were characterised by dislocation and change.

In addition to all this—affecting communities at a general level—it is necessary to take into account the points of articulation existing in each particular region. In Morelos articulation was defined by the symbiosis between Los Altos and the hot country, whereby the latter produced sugar cane and generally cash crops, as well as sugar and industrial derivatives, and the mountains provided cyclical cheap labour, grain to feed lowland inhabitants, and—on private estates—meat. In the lowlands, communal land was gradually absorbed by latifundia; furthermore, numerous settlements disappeared altogether (Hernández-Orive, 1973: 128). In the Highlands, the existence—and persistence—of the communal land has to be understood not merely in relation to the internal organisation of the villages, but, most important, in terms of the function of its produce: to send grain to the lowlands and to decrease the cost of labour. Following the same line of argument, other characteristics of the Highlands can be explained. For instance, lack of irrigation—and the consequent impossibility of working the land during the dry season—created a seasonal surplus of local labour; it also resulted in the possibility of using more water in the the lowlands. The use of the land almost exclusively for maize and pulses also seems to be related to demand for those products in the lowlands.[27]

If communal land could not have had a fixed 'function', the same applied to other institutions of local village life, such as religious festivies. By ignoring possible regional and diachronic variations, the functional approach to the *fiesta* system tends to be both teleological and extrapolative, i.e. to assume that certain observed consequences of the *fiesta* at a particular point of time and space—such as economic levelling and validation of political status—are also the reasons for the emergence of these ceremonies in all circumstances. As an alternative to functional interpretations, I suggest an approach in which all the social forces involved in the *fiesta*, as well as the relationships between these forces, are examined in detail.

The colonial fiestas

Religion and local government. The functioning of the politico-administrative framework created for rural communities required the presence of Spanish officers who enforced law and order, organised Indian labour and collected taxes (*tributos*). In these tasks they used the help of native officers. Initially, these were local traditional Indian chiefs (*tlatoani*, pl. *tlatoque*); but, soon after the establishment of the colonial system, in order to avoid concentration of power in a single person, the offices were allocated to men elected every year by the Indian members of the community, in the fashion of the municipal councils of Spain.[28] The top Indian officers were the *gobernador* (governor) and the *alcalde* (mayor). Other important officers were the *mayordomos*, who were directly responsible for the communal wealth (Gibson, 1967: 184). There were also a number of minor offices which varied according to the size and importance of the village; frequently these were occupied on a rotational basis by practically all adult males (Aguirre-Beltrán, 1953: 39; Gibson, 1967: 177, 186). In order to hold office at the top of the hierarchy it was necessary to have served in the inferior levels—what has been called 'the ladder system' (Carrasco, 1961). The Indians were encouraged to participate in the ladder system by the salaries allocated to the officers: as these salaries were drawn from the taxes, the officers had a special interest in making sure that taxes were collected efficiently. In addition, top offices were highly desirable because of the power, prestige and personal economic advantages implied in organising communal resources (Gibson, 1967: 188–9, 215). On the other hand, the functions of the officers were concerned not only with civil matters but also with the religious life of the community. Christian ideology not only justified the actions of the conquerors; it also gave the Indians a rationale for their submission to the conquest and the destruction of the old system (Mecham, 1966: 13). The presence of Spanish friars—Augustinian friars, in the case of the Morelos Highlands—and later of secular clergymen became essential to the maintenance of the colonial order. Missionaries and parish priests, in their labour of evangelisation, were in a closer contact with the Indians than the Spanish civil authorities (Gibson, 1967: 95). Missionaries frequently defended the Indians from ill-treatment by other Spaniards, which contributed to the acceptance of their teachings (Zavala and Miranda, 1954: 97). But the control of the Church representatives over the communities was not less stern than that of their civil colleagues, and for this control the priests also counted on the direct help of the Indian officers. These were in charge of gathering the people for mass on Sundays, of seeing that all adult persons in a community went to confession and communion at least once a year, of prosecuting pagan beliefs and practices, and of whipping those who did not fulfil their Christian duties. All officers were expected to be learned in basic Christian doctrine, and to be persons of deep piety.[29] The holders of minor offices, in addition, had to perform personal services to friars and/or other clergymen. It was common for the rise of a minor officer to the top of the ladder to be due to the recommendation of the parish priest (Gibson, 1967: 281); furthermore, the elections for major offices every year took place in the presence of the priest (Simpson, 1952: 91). It was therefore in the interest of the Church to support the smooth functioning of local administration because of the major role played by the latter in religious affairs, and because the salaries of the clergy were also drawn from the taxes collected in the community.

It is within this framework of colonial civil and religious administration that the emergence of *fiestas* has to be understood. The *fiestas*—which included High Mass, processions, meals, singing and dancing, bullfighting—were organised and financed by the Indians themselves, under the leadership of the native authorities. They provided the Indians with an occasion to demonstrate their piety and to attract the sympathy and protection of priests. But *fiestas* were in the first place introduced and promoted by the clergy for a variety of reasons. The images of the Christian saints—in whose honour *fiestas* were celebrated—were not only a convenient replacement of the old pagan deities in a society in transition from 'idolatry' to Christianity, but also provided a didactic opportunity to present the persons of the saints as examples of lives dedicated to the service of the Christian God. *Fiestas* were also the occasion in which religious plays organised by the missionaries were performed. These plays represented the life of different saints and scenes from the life and passion of Christ. In addition, other plays about episodes of the history of Spain and Christian Europe were also staged. The most popular of these plays, which can still be seen in many places throughout Mexico, portrayed a battle between Moors and Christians, in which, of course, Christians were the victors (Warman, 1972). The actors who played the role of Christian knights wore masks imitating European faces—with white skin and beard—and the play as a whole was an obvious reminder of the superiority of the Spaniards and their religion. A climax in the performance of the play was the appearance of the image of St James the Apostle, or Santiago, the patron saint of Spain, at whose sight the Moors fled in terror (Gibson, 1952: 38). The image of Santiago which is still venerated in the town of

Tlayacapan is a life-size wooden statue of the Apostle on a white horse, wearing a plumed helmet and armour, and brandishing a sword in his right hand. For the colonial Indians, it was probably difficult not to associate this statue with the figure of a triumphant Spanish warrior.

The barrio system in the colonial period. Symbols of power within a context of inequality, *fiestas* were also a tangible administrative mechanism. An essential aspect of their organisation was that participation and economic co-operation were compulsory for Indians, since they were linked to the process of allocation of houses and communal land. The group of Indians who joined together in the celebration of a particular *fiesta* in honour of a particular saint formed a 'brotherhood' or *cofradía* (Simpson, 1952: 71); membership of a *cofradía* frequently corresponded to membership of a ward or *barrio.*[30] *Barrios* were named territorial subdivisions formed by a set of households within a town or village, or else separate settlements subject to a principal town (Zavala and Miranda, 1954: 42). Each *barrio* had a chapel or shrine lodging the image of the saint whose name was adopted by the *barrio*; many chapels—and many images—still survive in the Morelos Highlands. In this region—and probably elsewhere—the formal functioning of the *barrio* and the worship of its saint were clearly defined. Members of the *barrio* were given the right to build and use a house within its territorial limits, as well as rights over a plot of agricultural land. In return, they had to pay a special *barrio* tax, which was especially directed to the financing of religious festivities. As house and plot were given to an adult male when he married, the system emphasised the weakness of kinship groupings other than nuclear families. On the other hand, the *barrio* system was also related to the distribution of authority, as the performance of minor tasks within the community's hierarchy of Indian offices was an obligation of all the adult males of each *barrio.*

I should mention that I am in disagreement with the treatment of colonial *barrios* by other writers. For instance, John Ingham has argued that a complex of twenty-six *barrios* existed in Tlayacapan and responded to 'a cultural model of the collective organisation of the clan-like groups known as *calpultin*' (1971: 615). He suggests that the Aztec *calpulli* (pl. *calpultin*) was virtually identical to the colonial *barrio*, and that even nowadays many traits of the *calpulli* persist in the community social organisation (*ibid.*: 617–21). In reference to 'Indian communities' in general, Aguirre-Beltrán has similarly confused the *calpulli* (kinship and territorial grouping) with the *barrio* (purely territorial grouping) (Aguirre-Beltrán, 1953: 40).[31] This point of view ignores the fact that the social context in which the institution of the *barrio* receives a meaning has throughout the years changed radically. During the colonial period Indians could, and did, move residence from one *barrio* to another, and therefore the relevant criterion of membership was not 'clan' affiliation but payment of *barrio* taxes. In addition, *barrio* affiliation was a means of counting and identifying those Indians who were needed for the seasonal labour force. (It is interesting to note that the *barrio* system was—and still is—rather underdeveloped in lowland villages, where labourers were totally incorporated into the plantation structure.)[32] After independence, *barrios* tended to disappear *qua* administrative mechanisms; thus membership was defined rather loosely in terms of residence.

Some of the main properties of the *barrio* system and its relationship to *fiestas* in colonial times can be seen in a dispute between a *mestizo* (half-caste) woman of Tlayacapan and the Indian mayor of the same town, as recorded in an eighteenth-century document.[33] This document reproduces the hearings which took place before the Spanish magistrate concerning the allocation and property of a house.

The case of the widow and the mayor

The woman (Josepha-María Vida) was the widow of a local Indian (Andrés Santa María); she and her husband lived in a house of the *barrio* of San Nicolás, to which he regularly paid taxes. The woman complained to the Spanish authorities because, after her husband's death, the mayor had intended to use his authority to give the house to another Indian. She alleged that the house was originally the private property of her husband's father (Diego Gregorio), and that the property was inherited by her husband and passed finally to her. In addition, she said that one of her sons, who was living with her, continued paying taxes to the *barrio*. Josepha-María was then using a double argument to assert her rights over the house; on the one hand, she claimed private ownership over it; on the other, she implied that even if the house belonged to the *barrio*, she was entitled to use it, as her son was a regular taxpayer.

The mayor (Joseph-Eleuterio Nopaltitla) replied that the house had never been the private property of anybody, and that Josepha-María did not have any legal document to prove it was hers. He pointed out that the woman was a *mestiza*—i.e. she was not even entitled to membership of an Indian *barrio*—but that she had not been threatened with eviction while her husband, who was an Indian and a member of the *barrio*, was still alive. The mayor also gave a different version of how the house passed to the woman's father-in-law, the Indian Diego Gregorio. According to the mayor's version, Diego Gregorio was a native of the *barrio* of Santa Ana who married an Indian girl in the *barrio* of San Nicolás and moved to live with his father-in-law. What was subsequently inherited by him was not the private property of the house, but the membership of the *barrio* with its inherent norms and obligations. Finally, the mayor said that throughout the dispute he had not acted on his own initiative, but at the request of the Indian officers of the *barrio*, who wanted to re-allocate the house, as none of Josepha-María's sons was willing to fulfil his obligations as a *barrio* member.

The mayor brought three witnesses to the magistrate in order to corroborate his own testimony. It should be noted that in his declaration he did not question the normative principles on which the woman's allegation was based, but simply stated that the facts produced by Josepha-María were distorted, i.e. that the house was not her private property but the property of the *barrio*, and that her sons were not paying taxes. The witnesses were Indian officers of the same *barrio* of San Nicolás, namely the *mayordomo*, the *caballo-pixqui* (the equerry in the stable of the parish priest) and the *merino* (the man in charge of carrying water and other 'supplies'—*raciones*—to the priests). They all confirmed what the mayor had said; they explicitly added that it was the custom for those who belonged to a *barrio* to co-operate in its *fiestas*, as well as to perform minor offices in the *república*, and that it was because of this co-operation that they were granted a house and a plot of land. The witnesses were particularly angry at Josepha-María's sons because they had refused to pay *barrio* taxes and perform any *barrio* duties; instead, they had tried to get housing in other *barrios*, and now, being sons of a *mestiza*, they had the pretension of owning a private house, as the *gente de razón* would do.

After the witnesses spoke, the court decided that Josepha-María and her sons could stay in the house on condition that the latter fulfilled all their *barrio* duties.

Fiestas and the formation of social fields. This legal dispute is relevant to the discussion of *fiestas*. The explanations of *fiestas* derived from the paradigm of the corporate community stressed a single purpose or function in their celebration; but I believe this is to ignore other social forces at work in the full context in which *fiestas* occurred. By social forces I mean the constraints which the behaviour of individuals or groups pursuing particular or collective interests put on the behaviour and expectations of other individuals and groups with whom they interact. The *fiesta* is an event occurring within a particular context in which social forces operate. The actual meeting of forces through interaction results in the formation of a field. The particular behavioural orientations and reactions of the people who interact in a field depend on their perception of what forces are present, and on their interpretation of how they are operating in that moment.[34] An event or situation is defined in several ways by different actors, according to the interests they themselves have and to their perceptions of the interests of other actors. Furthermore, the definition of a situation is also related to the actors' perception of the interests of individuals and groups who are not physically present but interact with the actors in other contexts.Finally, it should be stressed that perception is not a matter of mere subjectivity, but is subject to constraints derived from the actors' differential access to power and resources. This means that an actor's interests are perceived and taken into account by others in relation to his power.

In the analysis of colonial *fiestas* I want to show that they were open to different definitions and that these definitions were in turn conditioned by the position of the participants in a wider society which was far from being in a state of equilibrium.

At one level, the *fiesta* could be defined as a symbolic statement of the collective interests of a group of people whose rights over housing and land depended on the membership of a *barrio*. This membership was publicly acknowledged and reinforced through their common participation in the ceremony. At the same time, the boundaries of *fiesta* participation excluded non-members of the *barrio* from rights over the property of the communal land. This definition of the *fiesta* as a symbol of cohesion of the *barrio* members *vis-à-vis* strangers and potential intruders is implicit throughout the dispute between the *mestizo* woman and the mayor of Tlayacapan. But there are other levels of definition which also have to be taken into account. The cohesion, and the very existence, of the *barrio*, or of the Indian community, depended upon its integration and subordination to a wider framework of colonial administration; thus, in order to decide who were members of the community or *barrios*, it was necessary to depend on the verdict of the Spanish authorities. There are numerous documents which record the complaints of the Indians of the region against Spanish landowners who invaded communal lands and water, and it is clear that the Indians would be helpless against these threats without the protection of the Crown magistrates and missionaries. This protection was granted within a context of regional planning: destruction of communal land would jeopardise the convenient functioning of the division of labour between Highlands and lowlands (such protection was not granted to Indians in the valleys). The splendour of the *fiesta*, the eagerness of the people to participate in it, not only reflected the community wealth, devotion and cohesion, it was also a symbolic statement of the patterns of co-operation between the colony—in its twofold manifestation of royal bureaucrats and Church representatives—and the Indians.

In addition to these two collective forces—which we may broadly classify as created by the interests of the Indians over their communal resources, and by the alliances of interests between the Indians and the colonial government—attention must be drawn to the forces induced by the particular interests of individuals participating in the *fiesta*. The Indian officers, for instance, who were directly responsible for its organisation, occupied a position where opposing interests converged. They represented the interests of the Crown and the Spaniards, for whom they collected taxes and controlled the local population; but they also represented the interests of the Indians and were supposed to safeguard their resources. By succeeding in organising a good *fiesta* an officer was able to focus attention on himself and be recognised both as a loyal Christian subject of the Spanish empire, and as a skilled administrator of the wealth of his Indian fellows. But this did not prevent the officer from using communal resources to his own personal gain (Zorita, 1942: 159) or to the gain of his personal friends (this was hinted by Josepha-María); in that case a successful *fiesta* could serve as a means of distracting attention from an officer's corruption.

Fiestas were also opportunities for the officers to co-operate with the priests in their task of evangelisation and therefore to gain the support of the clergy in their rise up the political ladder. In this connection, we have to consider the different ways in which the situation could have been defined by the priests themselves. In chronicles of the period written by missionaries, the aspect of *fiestas* which is most emphasised is the great devotion of the Indians taking part in them (Simpson, 1952: 71 ff.; Benítez, 1965: 95–6). For the clergy this devotion was a confirmation of the success of their mission and of the importance of their role in an Indian village. This fundamental factor of the interests of individual clergymen in *fiestas* is related to their legal status *vis-à-vis* the civil colonial government. Powerful and rich though the Church was as an institution, the clergymen were in practice subordinate to the civil authorities. When America was discovered, it was the Crown and not the bishops or any religious order that assumed the responsibility for evangelisation of the realm. By virtue of the royal patronage of the Indies all ecclesiastical offices were to be filled by clergymen approved by the civil authorities, who also controlled the allocation of ecclesiastical salaries and tithes (Mecham, 1966: 26–7). Civil officers could also increase or decrease the number of labourers that priests needed for their personal service or to work in Church enterprises. On the other hand, a priest in charge of an Indian community could avoid total subordination to the local Spanish authorities by stressing the importance of his ideological role and by building up alliances with the Indians, particularly with the Indian officers, who were needed by the royal government in its everyday functioning. And *fiestas* were crucial occasions for expressing alliances between Indian officers and priests.

Last but not least, Highland *fiestas* were also symbolic statements of the symbiosis between villages and sugar *haciendas*. It was customary for the latter to help in the financing of festivities as a token of goodwill towards their seasonal labourers, or else as a payment (rather symbolic) for the use (occasional or permanent) of village pastures and woods. In Tlayacapan, for instance, the main festivity was San Juan's, whose image was—and still is—kept not in a *barrio* chapel but in the main church, as it represented the unity of all *barrios* within a single village. To this festivity the *hacienda* of Pantitlan contributed throughout the seventeenth and eighteenth centuries with honey, sugar and money, as well as with several bulls for a bullfight (Bonfil, 1971: 181–2).[35] Oral tradition maintains that many young mulattoes who worked in the *hacienda* came to this bullfight wearing colourful silk garments. A local dance, still performed in Tlayacapan at the Lent fair, represents the story of *hacienda* cowboys gathering cattle and taking it to a Highland *fiesta*.

Fiestas and the restoration of equilibrium. Even if *fiestas* may appear to have been crucial points of convergence for the interests of different categories of people, this does not necessarily mean that the communities *qua* communities were stable, equilibrated systems, nor that *fiestas* were mechanisms restoring a state of balance between interconnected parts whenever a disturbance occurred. The picture of the communities which emerges from documents of the period is one of heterogeneous categories of people in competition over resources. Indians competed among themselves and with non-Indians over houses, land and water; priests and royal bureaucrats competed for Indian labour; Indian officers were corrupted by Spanish landowners and conflicted with their communities over communal property; Indian officers competed among themselves for the favours of the Crown and the Church. I will mention only a few examples of legal disputes in the Highlands recorded in documents of the colonial period. In 1544 the Atlatlahucan Indians accused a Spanish landowner of using their cultivable land for grazing his cattle. In 1591 and 1592 the missionaries in Totolapan claimed rights over Indian labourers who had been allocated to work in the Chalco *repartimiento*. In 1591 the Tlayacapan Indians accused an Indian officer, who was allegedly protected by the priests, of embezzling taxes. Two *barrios* of Tlayacapan contended fiercely in 1592 over *jagüey* water, and the Spanish authorities ruled that *jagüeyes* should not be owned by a single *barrio* but by the *pueblo* as a whole. In 1601 the priests of the Oaxtepec hospital accused the Spanish magistrates of the Morelos Highlands and several other villages of not sending Indian labourers to work in the hospital's sugar mill, as was their duty. In 1604 a dispute over Indian labourers broke out between the priests of Tlayacapan and the Spanish *corregidor*. In 1629 a blind Indian of Atlatlahucan complained that, in spite of his physical handicap, the Indian officers forced him to perform *barrio* duties and to pay taxes. In 1746 the Indians of San Lucas (a settlement which no longer exists) and San Andrés fought over the use of water in a *jagüey*. In 1797 the Tlalnepantla Indians complained of the invasion of their communal land by Spanish landowners. In 1796 the Indians of Totolapan accused the Spanish authorities of preventing them from using firewood from the communal hills. In 1801 the Indians of Totolapan accused the Spanish magistrate and the Indian governor of selling *jagüey* water to outsiders.[36]

The celebration of a *fiesta* did not change this state of opposition among diverging interests, because opposition was a necessary consequence of the organisation of colonial society. Sometimes the *fiestas* themselves were occasions for violent fighting among members of different *barrios* asserting the superiority of their respective saints. On the other hand, *fiestas* provided a ceremonial expression of co-operative social relationships among the very same people who were divided by opposed interests in other situations. The ceremonies thus showed that the formation of fields of social action involving patterns of co-operation was possible in spite of the different ways in which the situation was defined by actors with diverse orientations. Even if stable equilibrium was never reached, even if opposition could not possibly disappear, these patterns of co-operation account in part for the persistence of the colonial order over time. However, they cannot be explained simply in terms of alliances resulting from the operation of local community forces; they have to be understood also in terms of the operation of forces in the wider society, as the goals, and the very nature, of any type of co-operative arrangement were in fact determined by the structure imposed by the colony.

Dialectical processes of conflict and alliance in the colonial society implied the participation of individuals and groups having well-defined positions within a single system. These positions or statuses were expressed in terms of ethnic categories, of territorial rights, and of rank within a hierarchy of offices. However, the status of people like Josepha-

María was problematic. She was a *mestiza*, and as such she did not properly belong to either parental group. The only way in which she could claim membership of the Indian community was through marriage to an Indian, or through attachment to her sons (who could claim membership of their father's ethnic group if they were willing to carry out obligations such as tributes and co-operation in *fiestas*). On the other hand, *mestizos* could not enjoy the privileges of the Spanish group (Hansen, 1971: 137). A similar situation was faced by the creoles (*criollos:* people of European blood born in New Spain), who were also excluded from important political posts. Some creoles and *mestizos* had access to wealth—through inheritance, or through the use of their wits—others joined the army or the Church. But a substantial proportion of them had to live on the fringes of the society.

The end of the colonial system

A useful tool for the understanding of dialectical processes is Max Gluckman's distinction between rebellions and revolutions. A rebellion is the result of conflicts which can be settled within the working principles of the existing social system. A revolution is the result of irresolvable contradictions. *Contradictions*, in Gluckman's terminology, differ from *conflicts* in so far as the former cannot be solved without changing the nature of the system in which they occur (Gluckman, 1955, 1963, 1965). However, certain conflicts can in certain circumstances become contradictions, and thus a series of rebellions may result in a revolution. The oppositions described above are to be classified as conflicts; but one important cause of the breakdown of the colonial order was the accumulation of irreconcilable oppositions in the social system, from the level of village organisation to the highest levels of national authority. The government could not adequately protect the interests of the Indian population while at the same time protecting the penetration and dominance of the Europeans in all economic, political and religious activities. The growing sector of *mestizos* and creoles, who did not find a proper place in the colonial structure, eventually turned against the Crown and provided the leadership of the 1810 movement of independence. The *hacendados* (Spanish or creole), who monopolised grain and caused artificial scarcity and rampant inflation, could not possibly reconcile their interests with those of the general public, or even with those of the merchants and the magistrates—with whom they often quarrelled (Florescano, 1971: 120–1, 211). Within the Church, there were bitter rivalries between regular and secular clergymen, as well as between the high-ranking churchmen—mostly Spanish—and the lower clergy—who were creole and *mestizo* (Simpson, 1952: 77–81). When the independence movement started, its leaders were two rural clergymen, Hidalgo—creole—and Morelos—mestizo—, who not only attacked the privileges of the Spaniards but also advocated the inalienability of community land and the fragmentation of the big estates.

All these cleavages were present during the whole colonial period; but it was in the second half of the eighteenth century, when new administrative measures were imposed by the Ministers of the Bourbon kings, that they became impossible to bridge by cross-cutting alliances. Two measures particularly affected the economy of Morelos. The first was the heavy taxation imposed by the Bourbon administration on local economic enterprises (Brading, 1975: 46 ff.). The largest land holding in the Morelos Highlands, the *hacienda* of San Diego Huixtla, which covered 3,147 hectares of pastures and forests, was sold by its owner, the archbishopric of Mexico, to the Atlatlahucan Indians in 1779.[37]

The fact that the Atlatlahucan Indians were able, at least initially, to pay the price of the *hacienda*, poses the question of where the Indians got the money from. The answer is related to another administrative measure: the abolition of *alcaldes-mayores* and *corregidores*—i.e. those royal officers who controlled the allocation of Indian workers at the local level (Florescano and Gil, 1976: 212 ff.). This made it possible for the Indians to sell their labour in the free market. African slaves were no longer imported into New Spain, so that sugar production and expansion relied heavily on Indian labour (Humboldt, 1953: 186). Until the Bourbon reforms, planters had enjoyed official privileges and used extremely cheap labour from the communities, in spite of the protection granted by the Crown to the Indians (Hernández-Orive, 1973: 135); but under these new circumstances they were forced to offer comparably higher wages if they wanted to compete with the increasing demand for labour from silver mining, which was booming at the time. To make things worse (for the planters), the measure also meant that grain and agricultural products were no longer controlled by local authorities, and could be sold by the Indians wherever they pleased. And this happened during a succession of disastrous summers which ruined several crops and resulted in a sharp rise in the price of grain (Florescano and Gil, 1976: 280–1).

On the other hand, planters adopted strategies designed to reduce their costs by renewing their attack on village land and water, in both the Highland and the lowland areas, as has been documented by Sotelo-Inclán (1970) and Hernández-Orive (1973). It should be remembered that sugar-cane fields required heavy investment: planters used this as justification for considering a field their property once the expenditure on irrigating it had been laid out. But at the end of the eighteenth century the local population had—as in other parts of New Spain—begun to grow, and in many places villagers claimed their land back. The popular rebellions led by Miguel Hidalgo (1810–11) and Jose María Morelos (1811–15)—after whom the state of Morelos was to be named several decades later—found strong support among villagers and labourers of the Cuautla-Yautepec region.[38]

Haciendas, ethnicity, power

Prior to the discussion of nineteenth-century regional changes, it is pertinent further to examine two analytical points which are essential for the understanding of Morelos rural life: the relationship between villages and haciendas, and the nature of ethnic labels.

My view of the impact of *haciendas* on village life is sharply different from the approach which sees the relationship between large land holdings and Indian communities as reinforcing internal homogeneity, social boundaries and 'corporate' structures in rural settlements (Wolf, 1965: 99 ff.). I have the following misgivings. In the first place, it presupposes that *hacendados* were interested in rent and not in intensifying production for the market. Morelos *haciendas* do not fit into this 'feudal' pattern. Secondly, it assumes that all villagers had the same type of link with the *hacienda*, without paying attention to complexity in *hacienda* organisation, which demanded very different sorts of participation on the part of villagers and hence created a criterion for social differentiation in communities (Carrasco, 1957: 17). Thus the Highland people could relate to *haciendas* not only as seasonal labourers in sugar-cane fields. They could monopolise or accumulate grain and sell it to the valleys—even in times of scarcity, if local authorities had a part in the business. They could also trade the commodities produced in the Highlands: leather, wax, meat, pottery, silver, fruits. The production of these things was stimulated by the concentration of people and money brought about by *haciendas*.

One could say that all these activities were performed by 'non-Indians', and that the purity of the 'Indian community' was not touched by outside contacts. This is difficult to prove. On the other hand, what did it mean to be an Indian? The persisting clear-cut *verbal* distinction between *naturales* and *gente de razón* did not merely emerge from differences in race or cultural background among a given set of people, but rather from national legal arrangements for the allocation of crucial resources: power, authority, wealth, and particularly control over labour and landed property (Zavala and Miranda, 1954: 31). The distinction between social categories of people with unequal access to resources was expressed primarily in terms of ethnicity. However, people could—and did—change their ethnic label. Poor *mestizos* (like Josepha-María) could be grafted on to an Indian *barrio* so they could receive a plot of land, thus becoming Indians. In turn, Indians could become *mestizos* or even *criollos.*[39] To this effect, genealogies were falsified by paying a bribe to actuaries and authorities (Mendizábal, 1974: 6). In the Highlands several Nahuatl surnames were translated into Spanish in this type of arrangement. (Tepanohaya became Pedraza; Xochitl became Flores; Nopaltitla beame Campos.) The multiplicity of shades of colour ('castes') resulting from frequent racial mixture facilitated ethnic manipulation. On the other hand, the population of the Highlands, through active participation in a complex economy, became familiar with the Spanish language. A document of 1808, recording a land dispute, certifies that the Indians of Totolapan did not need an interpreter, 'because they were clever [*ladinos*] enough in the Castilian language'.[40] One could reasonably assume that they were also *ladinos* in affecting other non-Indian manners.

One may thus speak of mechanisms of 'situational selection' by which people could manipulate norms and participate at different levels of society. However, individual manipulation—as shown in the analysis of *fiestas*—had clear limits. An individual could not help taking into account the existence of ethnic labels, or the existence of a common frame within which resources were organised and allocated. It is pertinent to refer to the existence of an overall power domain during the colonial period. Power was effectively vested in the King of Spain's representatives: the Viceroy and the Audiencia. They controlled the Indians' environment through *barrio* and communal land regulations, the entrepreneurs' through patterns of labour allocation, even the Church's through patronage and control over tithes and Indian labour. In a way, the Spanish Crown destroyed the foundations of its own power domain by reforming the local administrative system in an attempt to become more flexible and efficient. The Bourbon reforms opened the way for all accumulated contradictions to become violently manifest. The violence released would not settle until a new power domain was created.

The nineteenth-century crisis

The nineteenth century in Mexico has often been conceived of as a chaotic break with the colonial system and a long prelude to the twentieth-century revolution. In this conception, an important motif in the prelude is the process of village land expropriation which culminated in the application of disentailment laws in the second half of the century.

In Morelos, it is possible to distinguish two stages in the decrease of village land and the increase in the size of private estates. The first lasts until about 1880. During this first period the process of communal land expropriation took place mainly in the lowlands. In spite of a critical scarcity of capital in the country as a whole, lowland sugar-cane fields and refineries never stopped production for the national market and even for export. In turn, Highland communities provided their supply of cereals, meat and labour (Sotelo-Inclán, 1970: 225). The regional division of labour persisted, but in a less peaceful context. In addition to conflicts and contradictions related to political upheavals at the national level, there were specific problems in Morelos derived both from land disputes and from the resistance of landowners to paying their labourers adequate wages (as had been the case since the eighteenth century). An additional source of trouble was the rivalry between Spaniards, creoles and *mestizos* seeking to control local political posts. Spaniards were finally excluded from public offices; but contradictions between *hacienda* and village, capital and labour, never ended, and had cumulative effects which came into the open in numerous acts of violence (Sotelo-Inclán, 1970: chs. XVIII–XX; Warman, 1976: 90–3). At village level, two phenomena seem particularly important: the transformation of local government and the reorganisation of the *fiesta* system.

The second period (1880–1910) was characterised by a sudden expansion of the world sugar market, which led to new growth for the Morelos refineries and, correspondingly, a need to directly monopolise more land. Lowland villages lost practically all their terrains, and *hacienda* voracity started to engulf also the lands of the mountain communities. This second period immediately preceded the revolution.

Changes in the municipalities. The restoration of the municipal system put every town, together with its satellite villages, in the hands of a council of 'prominent citizens'.[41] These local authorities were no longer Spaniards but resident *mestizos* or creoles, such as cattle-raisers, merchants and workshop owners in Highland villages. The municipalities had no clearly-defined functions, so the activities of the local authorities became more than ever oriented towards personal profit rather than administration (Aguirre-Beltrán, 1953: 58–61; González, 1956b: 149). On the other hand, the Indians (i.e. the holders of communal land) had been made 'equal citizens' before the law—although ethnic distinctions persisted in practice, as the system of land tenure associated with ethnic labels did not change—and were no longer subject to any kind of personal tribute. As the new elite could not fill their pockets with Indian taxes, a manner of enrichment not altogether infrequent was forceful appropriation of village land (Sotelo-Inclán, 1970: ch. XXII). This process was encouraged by the fact that only big villages (over four thousand people, in Morelos) could become municipal heads; and only heads could hold title to communal land. In other words, the rights of smallish settlements over communal land now existed merely as a concession of the municipal council. In Morelos the ninety-four *cabeceras* of 1800 became twenty-two municipalities after independence (Gerhard, 1975: 576). In the Highlands, Tlayacapan was the only remaining municipality until the middle of the nineteenth century. In practice, municipal councillors could redefine the boundaries of village or *barrio* plots and sell (or take over) those pieces of territory which they considered were not needed by the villagers.

A double crisis: Church and community. On the other hand, the financing of religious festivities, which had been closely associated with tax collection, was no longer formally obligatory for members of a community or *barrio*, and could not be enforced by law. A document of 1862 records that a man of Atlatlahucan demanded protection from the municipal authorities because the neighbours of his *barrio* wanted him to pay a share of the expenses of a *fiesta*. The authorities confirmed that nobody should be compelled to give money for a religious festivity, as it was a matter of devotion and not of obligation.[42] This verdict is diametrically opposed to the decision made by the Spanish court in the legal dispute described above, which had occurred 100 years earlier.

The *fiesta* system was also affected by a substantial decrease in the number of the clergy, and by changes in the political role of the Church.

The decline in the number of clergymen had begun during the Bourbon period, when the religious orders were stopped from holding ecclesiastical offices. After independence, many Spanish priests left the country; in 1830 the total number of clergymen had been reduced by two-thirds.[43] Many villages did not have a resident priest for long periods of time.

At the same time, the liberal sector within the Republican government launched a campaign against the institutional power of the Church. This power was based on the official character of Catholicism, and on economic dominance: the Church held at least a quarter of the national wealth. The government's freedom of action was limited by the Church's opposition to any changes which could threaten clerical privileges (Mecham, 1966: 354). The liberal faction finally defeated its conservative opponents, and issued the Reform Laws—promulgated in 1859, but not fully enforced until 1873—which separated Church and State, and dealt a heavy blow to ecclesiastical finances (Quirk 1964: 63 ff.).

The Reform Laws disentailed not only ecclesiastical property, but also the land held by Indian communities. Thus the material basis for the *barrio* system ceased to exist. The formal object was to give the descendants of the Indians the opportunity to acquire individual property so that they might participate in a free capitalist economy. What happened was that communal land was often acquired by the economic and political elite of each community, or by outsiders (González, 1956b: 316–20). Many descendants of the Indians were now landless and had to work full time as sharecroppers and labourers both for local landowners and for the sugar *haciendas* of the lowlands. In the Highlands, many poor families managed to retain a plot, especially terraces in the hills: but now they did so as private owners. The image of the *barrio* saint was no longer the symbol of the formal rights of a group over communal resources. Many *barrio* chapels were abandoned.

Folk religion. However, not all the *fiestas* disappeared. Their persistence has to be explained in relation to the new social context of the communities, unless one assumes the existence of forces of cultural inertia. In this respect a distinction has to be drawn between the religion of the *mestizo* elite and the *fiestas* of the lower strata of the population.

One important aspect of *fiestas* among the lower strata was that the presence of the priest was no longer essential to the performance of a ceremony. This was made possible by the development of a category of lay ritual specialists in the communities, who led collective prayers, hymn-singing and music.[44] The sporadic visits of the priest to a locality sometimes coincided with the celebration of a *fiesta*; in that case, the priest was invited to celebrate a mass for the saint; but other ritual activities within the *fiesta*—such as dances, processions, prayers—were still led by the ritual specialists. The result was the emergence of what has been termed 'folk Catholicism' (Carrasco, 1957; Madsen, 1967), which to a large extent functioned—and still functions—separately from the official religion of the Church. During the colonial period the Church had tolerated—even promoted—some form of religious syncretism; but then it had the power to control beliefs and practices, and to punish deviation from orthodoxy. Now it had lost its power. Priests were often reduced almost to the status of employees of the people, who paid them a fee for their services. As they needed the fee, having lost much of their wealth and their rights to tithes, priests had to accept the patterns of local custom in ritual, even if they were somewhat heterodox (González, 1956 a: 254; Carrasco, 1957: 32).

Folk religion, on the other hand, was related to the opposition between the local elite and the poor descendants of the Indians. The *mestizos*—after the *fiestas* had lost their economic importance because of disentailment—tended to adopt the religious practices of urban groups. For the elite, religion was a personal affair, associated with inner experience rather than with collective symbols. For the people, religion was interwoven with many other spheres of social life. Their economic activities had traditionally included small-scale trade; and trade was closely associated with the *fiesta* cycle. *Fiestas* were occasions on which vendors from different villages found a large number of people ready to buy new clothes and food for a set of ritual activities which were at the same time their main recreation. In Tlayacapan potters produced the incensories (*sahumerios*) and candlesticks used in the offerings to the saints. In many other *pueblos* of the Morelos Highlands women made decorative items for the chapels—such as elaborately ornated candles (*escamadas*), paper flowers, and ribbons.

These recreational and commercial aspects of *fiestas* have existed since colonial times. They persisted—and moreover became an expression of dominant social forces—in so far as the sector of the population classified as 'Indian' continued to be excluded from participation in the wealth and status symbols of the upper strata. The Morelos villages which legally lost communal property—because they were not municipal *cabeceras* or, later, owing to disentailment norms—could sometimes use their lay ritual specialists and *fiesta* organisers to keep at least some village land (Gerhard, 1975: 577). Such people were represented formally as private owners of plots which in fact were administered and used by a whole group, in their very existence defying, and expressing their opposition to, the standing order.

But, again, we find alliances as well as opposition in the social context of the *fiestas*. These events were partly financed by the elite itself, as a strategy for maintaining good relations with their labourers and sharecroppers. Tlayacapan bullfights continued to be paid for by lowland *haciendas*, and Atlatlahucan merchants paid for candles and drinks during St Matthew's festivities. Folk religion was not a mere expression of a static folk culture which responded to 'local needs' (Carrasco, 1957: 8). Like any social institution, religion was immersed in complex processes which took place on a dimension wider than village or even regional social life.

Progress and rebellion, 1880–1910

It is, perhaps, advisable to stress that terms such as 'Indian sector' or 'lower strata' should not be understood as denoting homogeneous social groups with a consistent organisation of their own. This also applies to other social categories to be found in nineteenth-century Mexico: small ranchers, sharecroppers, labourers, muleteers, peddlers, soldiers . . . and bandits (González, 1956a: 332). None of these categories described an organised group (or a permanent set of groups), but a loose aggregate of people whose status and behavioural expectations were conditioned more by an encompassing national society than by any local micro-structure. The precarious social order was not a function of balance within a village or among the members of a single sector; it was the result of a balance of social forces in heterogeneous fields. Morelos organisation mixed *hacienda* land and capital with seasonal labour in enterprises which played an important role in the national economy. To be a *villager* was to be a member of this large organisation, symbolised in the equally complex dimensions of the *fiesta* system. Although decadent and shattered, this system continued to reflect principles of opposition and alliance in a society of unequals. However, organisational principles were radically modified after 1880 in such a way that oppositions became stronger than alliances.

New economic strategies. The process of disentailment accelerated under Porfirio Díaz's dictatorship (1876–1910). At the national level, the rights over land were transferred on a gigantic scale.[45] The process was consistent with an overall 'modernising' economic policy which had as its main goals production for the world market, the creation of a vital network of railways and roads, and the establishment of European-style industries.[46] Morelos fitted into this framework quite well: the collapse of sugar-growing in Brazil caused a boom in demand on the international market and conditioned the emergence of new strategies on the part of lowland *haciendas*. One of the first was to modernise technology by importing heavy machinery and renovating the old-fashioned mills. This could be done without major problems, thanks to the railway: backed by the political and economic weight of the *hacendados*,[47] it reached Cuautla in 1881 and Yautepec in 1883, only ten years after the first line to Veracruz had been completed. Other strategies were the securing of more land, water and labour. Land and water had to be grabbed from villages in order to expand production. Labour was apparently available: the state population (66,000 in 1800) had grown from about 120,000 in 1860 to about 155,000 in 1880, and reached 180,000 in 1910 (González, 1956a: 112, 120), partly because of the attraction of expanding enterprises to migrants. 'Plantations in Morelos became company towns, their permanent population ranging from 250 to almost 3,000' (Womack, 1969: 43). Wages were kept low, and soon local people and migrants started looking for land to grow their own food—but land was monopolised by the *haciendas*. Eighteen villages—more than anywhere else in the country—were destroyed between 1884 and 1905 by sugar expansion: not only were the fields taken, but the housing areas too. Families were mobilised as permanent workers. The landscape changed abruptly in many places: the town of Yautepec—according to the testimony of the novelist Ignacio M. Altamirano—was surrounded by orchards of orange and lemon trees in the 1870s which later disappeared completely.[48]

At the turn of the century Morelos had two contrasting faces. It was (in 1908), after Hawaii and Puerto Rico, the main sugar-producing region in the world, boasting a yearly output of over 52,000 metric tons (Womack, 1969: 49). At the same time, the voracity of plantations for land to grow sugar cane created an acute shortage of maize and other foodstuffs for consumption (Holt-Buttner, 1962: 15), which in turn raised the cost of living for workers. This fact had important repercussions for Highlanders. Their land could not be used in the direct production of sugar cane because of climatic conditions and lack of irrigation; yet land use in the mountains became more than ever dependent on the growing demand for meat, vegetables and cereals for the lowland *haciendas*, so much so that even local Highland consumption of maize had to be reduced. Demand for wood for fuel also increased—and eventually led to the devastation of hills and forest land. These demands could no longer be met by the existing organisation of regional production, i.e. that which left the direct control of seasonal land and forests in the hands of Highlanders—or so the *hacendados* thought. It was necessary to control production and distribution more directly lest a part of the product never reach the market. Thus several lowland *haciendas* bought tracts of land in the Highland *muncipios*: San Carlos, Buenavista, Oacalco expanded from the south. From the north-east, Mayorazgo and Xalpa, two *haciendas* in the state of Mexico, took over pasture land in Tlalnepantla and Totolapan to herd their livestock. In addition, two *haciendas* flourished in the Highlands themselves: Coatepec (which consisted of more than 2,000 hectares) and San Diego Huixtla (Tepantongo, more than 3,000).[49] These local *haciendas*, and the smaller *ranchos*, which also increased in size (they ranged from 50 to 300 hectares) and number at the time (see De la Peña, 1980: Table 3.2), functioned solely to supply lowland demands.[50] (Some *ranchos* were also owned by the plantations; for instance Alotepec, in the *municipio* of Totolapan, was for several decades the property of *hacienda* San Carlos; Mancera and Jagüeycillo, in the *municipio* of Atlatlahucan, belonged to *hacienda* Coahuixtla.) As the railway crossed the Highlands southwards, the export of produce—and labour—was easy and fast.[51]

Politics and disorder. In 1869 Morelos had become a separate federal entity; previously it had been a section of the vast state of Mexico. Planters were able from the moment of birth of the new state to exert influence over its government. This influence—which also reached spheres of federal power—was directed towards obtaining economic advantages for their enterprises: tax exemption, protective laws (sugar imports into the country were heavily taxed, thus allowing domestic prices to rise and diminishing competition), loans and transport (railway) (Womack, 1969: 42, 49, 53). But the planters were also extremely interested in maintaining law and order on their side. From 1894 to 1908 the office of Governor was held by Manuel Alarcón, an old soldier who shrewdly managed to contain peasant outbursts by making promises of land restitution and even, occasionally, defending the rights of villagers against sugar *haciendas* (Womack, 1969: 15). His successor, Pablo

Escandón (the owner of Atlihuayan, a prosperous modern mill near Yautepec), paid no heed to popular discontent. More than ever the district prefects sought to repress the frequent outbreaks of rebellion: rebels were forcibly enlisted in the army, or sent to work in labour camps. Old people in the Highlands still remember stories of the authoritarian rule of the Yautepec prefect (Yautepec being the district head for the Highlands) and his bailiffs.

The label of bandit was often applied to all sorts of people who expressed active dissent from the *status quo*. The elegant *plateados* of Morelos—the 'silver men', perhaps the most famous bandits of late nineteenth-century Mexico, whose feats and felonies are described by Altamirano in his novel *El Zarco* (1888)—raided villages and *haciendas* (burning down the *casco* or manor house of Pantitlán[52] as well as the biggest candle workshop in Tlayacapan) with the object of getting spoils and not of changing the existing order. Yet these rebels (in Gluckman's sense) did not share a common attitude to the society they fought against. They were recruited from multiple points of the social spectrum: from idle soldiery who had not benefited in the many coups and power reshufflings before Díaz's regime; from runaway *hacienda* labourers; from army deserters; from villagers who had lost their land; from small ranchers whose economy had been disrupted and displaced by the hegemony of the plantations. If many of these people had as their main interest merely to partake of the lion's share, others had as an aim the very destruction of a system in which social cleavages were practically impossible to bridge. Banditry, in this sense, was in Morelos a direct ancestor of the Zapata revolution, in so far as both emerged from the same social context of sharply opposed interests.

The revolution

'Of the hundred *pueblos* in the state in 1910, there was probably not one that was not involved in a freshly embittered dispute with a neighbouring *hacienda*' (Womack, 1969: 61).[53] Yet the Morelos revolution was not a mere 'peasant movement': it responded to the discontent of heterogeneous groupings and categories. A significant sector of city and town dwellers—teachers, journalists, lawyers, doctors, petty traders—felt increasingly uncomfortable with Díaz's dictatorship, with their own exclusion from the praxis of politics, with the planters' predominance and privileges.[54] As in other parts of the country, local small-scale industry and trade were affected by the availability of products brought by the railway. The region of Cuautla had been the theatre of generalised rebellion and repression at the time of the State government election. It is therefore significant that the leading figures in the Zapatista movement were not only agrarian fighters but also people from wider social origins, with wider ideologies and interests. Zapata himself was not a peasant but a horse dealer. Pablo Torres Burgos and Otilio Montaño were schoolteachers. Felipe Neri was an industrial sugar-mill worker. Manuel Palafox was a student turned salesman and accountant, Gildardo Magaña an ex-seminary student, Antonio Díaz Soto y Gama a leftist intellectual.

How then did the Zapata movement come into existence? It is important to note, first, that it was the result of the gradual coalition of several independent uprisings, which operated not only in Morelos but also in the states of Guerrero, Mexico and Puebla, and which had started for different reasons: land disputes, repression, armed rebellion against Díaz's dictatorship and in favour of the presidential candidate, Madero. Second, this coalition can be understood only in the context of a national political and social revolution. Third, the idea of land reform was clearly formulated and adopted as an overall banner only when the movement was already afoot.

The emergence of Zapata. In 1909 Emiliano Zapata, a young, dapper horseman of the village of Anenecuilco (near Cuautla), was elected by his fellow villagers as their representative in a legal quarrel against the *hacendados* who had taken village land. He soon decided to resort to direct action: having organised a group of eighty villagers armed with old guns, hoes and *machetes*, he took over the plots which the hospital *hacienda* had invaded. People from neighbouring villages came to him for advice, and he ruled that the old colonial land titles should be unearthed as a basis for marking the rightful boundaries of village land, and that invaders should be expelled. It was at this time that Francisco I. Madero[55] published his *Plan de San Luis Potosí*, calling for changes in the political and social structure of the country and urging an armed national uprising. Zapata joined him because of the clause advocating rural justice. The *Plan de San Luis* began to fuse the Morelos guerrillas into a single, if not yet very well co-ordinated, force—and their first collective action was a victory: the siege and capture of Cuautla, which was apparently abandoned by the Federal army—busy as it was fighting rebels in many other parts of the country. Madero's forces seemed to be growing stronger everywhere, so Díaz resigned and left the country in May 1911. From Cuautla, Zapata sent orders that the villagers should claim the land—although he made it clear that the plantations should be left in peace with the land which was rightfully theirs. The wave of land seizure that followed, and the alarm and complaints of the *hacendados* and their friends—still an important social and political group in Mexico—led Madero himself to think that the Morelos rebels should be restrained. But his attempts to negotiate with Zapata and other local chiefs failed. The Federal army occupied Morelos, expelling people from the fields. In November 1911 Zapata, now back in guerrilla action, issued the *Plan de Ayala*, a manifesto repudiating Madero, who was now President, and putting the agrarian problem as a priority for the revolution—as important as the institution of political democracy.

The *Plan de Ayala*, like the Cuautla siege earlier, gave Zapata a nation-wide reputation. Newspapers now referred to 'the Zapata army', although he did not become effective head of the movement until 1913, nor was there a proper army—rather a series of rioting bands. Slowly the common point of reference for all these bands became the *Plan de Ayala*. Zapata, in turn, was conscious of the need for discipline, organisation and clear aims in order to create a truly revolutionary movement and to refute the accusation of banditry which the press, the Federal army and the planters levelled at all Morelos insurgents.

In February 1913 Madero was overthrown by a military coup and murdered. The President was now General Victoriano Huerta, previously a leading figure in the military repression of Morelos villagers. He announced a campaign against 'bandits' and the full restoration of plantations and *haciendas*. There followed a year of civil war. Federal action—killing, burning, conscripting people into the army by force—contributed to unify (in mind if not in practice) practically all villagers under Zapata's leadership. And the latter, together with other local chiefs, formed the Revolutionary *Junta* (Council) of the Centre and South of the Republic, of which he was elected president. The *Junta* readjusted the *Plan de Ayala* to the new circumstances and, more important, started putting it into practice. Throughout the territory dominated by the Morelos revolutionary forces, land was distributed according to the old village titles—if they were available; if not, the elders were consulted. The Federal army made little progress: the whole nation was in arms, enraged by Madero's assassination. A national, 'constitutionalist', revolutionary movement was taking shape against the Federal forces. Meanwhile Zapata's influence extended over a substantial area of the states of Guerrero and Puebla. In June 1914 the whole of Morelos was his. In August the constitutionalists controlled the rest of the country. Huerta fled to Europe.

But, in all these years, what had happened in the region of the Morelos Highlands? A useful way of answering the question is by examining the life of a Tlayacapan villager. The revolution was not merely a series of glorious feats but a real change in local society, the impact of which can only be fully appreciated through the concrete dimensions of human lives.

A witness: Rosendo Rojas[56]

Rosendo, the son of Narciso Rojas and Mercedes Vega, was the descendant of an old Highland family: his parents, grandparents and great-great-grandparents had been born in Tlayacapan. His father was poor, but believed in the value of education; thus Rosendo—who was born in 1875—and his sister Antonia were sent to the municipal teacher, a woman whose fee was paid by the local authorities, in order to learn to read, write and count. Antonia made quick progress in literacy, and later became a private teacher: she gave elementary lessons in her house. Rosendo, in his turn, started farming the land with his father—a peasant farmer—as soon as he was old enough for the task. But he always thought education was an asset—it could at least help a countryman to avoid being cheated by city folk.

Heliodora Santamaría Toscano, the girl Rosendo took as wife, never went to school, nor did she learn how to handle written symbols; but she was beautiful and hard-working and her husband was very happy with her. When they married she was fifteen years old, and Rosendo twenty-five. They had eleven children; six died in infancy. In the first and second years of his marriage Rosendo continued working for his father, but soon he had to supplement his income as a seasonal labourer in the lowland sugar-cane fields. He thus went to many *haciendas*: Oacalco, Cocoyoc, Casasano—even to the more distant San Pablo Hidalgo, Calderón, Treinta. The demand for cane cutters was high—but the wages were low.

Rosendo's father, Narciso, worked two small plots of his property. One of these was far away, in a dell behind the Cerro de la Ventanilla.[57] It drew a little water from a source called Chautzingo, and mangoes could be grown with some profit. The other plot, in the plains south of the town, was used to cultivate chillies seasonally, for chillies had a good market. Both plots together amounted to less than two hectares—not much to be shared by Narciso and his three male sons (Rosendo, Cándido and Eduardo), especially when the three were married and had their own families. There was not much chance of getting more land of good quality. The best local plots—many of those which had previously been communal property—were in the hands of the local rich and the *hacienda* owners. What was available was hilly terrain of little use. So one day (around 1905) Rosendo decided to leave his father and look for more permanent work in the valley.

A full-time job was not easy to get. Rosendo found employment as a *tariero*: he had to do specific *tareas* (chores, tasks) like sowing, weeding, carrying loads, for which he received a *jornal* or wage of half a *real*.[58] If the day's work was particularly heavy he could earn a full *real*. The *tarieros* who came to the *haciendas* from the Highlands, or from other distant places, usually arrived on Monday morning and returned home on Saturday evening to spend a short weekend with their families. To come and go every day was tiring, as the journey had often to be made on foot. There was the train; but it did not reach all the places where the workers lived and its timetable was not regular—it was adjusted to the needs of transporting sugar or other products. Food was sent every day by their wives: tortillas, chillies, some fruit, occasionally *cecina* (dried meat). The *hacienda* paid *tlacualeros* or errand boys who fetched the food, but only for the *tarieros* who were actually at work on a task. The *tarieros* who had not been given a job to do that day had to pay the errand boy from their own purse. [59]

Searching for more security, Rosendo managed to become a *gañán* or permanent worker in Oacalco and later in San Carlos. He now received an *apero*—plough and oxen—and a daily wage of half a *real*. The oxen were not very good and the work was hard, but the regular pay was enough for survival.

Life was not that hard for all in Tlayacapan. There were the few rich who monopolised land, and also a handful of merchants, pawnbrokers and moneylenders. The lands of the wealthy grazed cattle or were used to grow tomatoes, plums and chillies, as well as maize and beans. The main market for these products was in Cuautla and Cuernavaca, but buyers from Milpa Alta, Xochimilco and even Texcoco came to Tlayacapan periodically. Heliodora, Rosendo's wife, decided to participate in local commercial life. She bought coffee—grown in domestic back yards—ground it, and sold it in the central square on Wednesdays, when the open market was traditionally held and there was a big influx of outsiders. The business prospered, so she set up a permanent stall. It even brought in enough to buy a small piece of land on the outskirts of the town which belonged to an uncle of hers who was childless and getting too old for farming.

In 1911 two names caught the attention of the villagers: Madero and Zapata. Madero was a symbol of the rejection of Díaz's regime. A few farmers and small shopkeepers founded an 'anti-re-electionist club', which did not meet openly lest the Yautepec bailiffs took it seriously. In any case, the members of the club—Rosendo was one of them—did not know exactly what to do. With Zapata, things were different: his agrarian programme immediately made people take a stand for or against him. The rich said he was a mere bandit. When he took Cuautla several villagers went to join him. The rich panicked: they knew they were not loved by their neighbours. Their accusations had sent several people into the army or to jail. The very day that Díaz's resignation was announced, Albino Salazar, a cattle-raiser and pawnbroker, was shot in broad daylight. One by one the proprietors fled the town by night, taking what belongings they could carry. The abandoned land was taken over by the poorest people.

But most neighbours were scared. From 1912 to 1914 the atmosphere was tense with mistrust and anguish. Even if most people sympathised with Zapata, only a handful actually joined the movement. Two men, Feliciano Polanco and Miguel Alarcon died gloriously in battle, and were buried in the monastery yard. On the other hand, some people disliked *zapatistas*. There was a revolutionary officer called Nabia who became friendly with Rosendo and Heliodora, and visited their house frequently, whenever the *zapatistas* passed by. The Federal soldiers also visited Tlayacapan, though they never committed atrocities, as they did elsewhere. Yet a woman denounced the Rojas for being rebels, and they were almost caught for it. When the *zapatistas* came back the woman was hung by the hands in a tree as a public lesson.

Food was scarce, and there was no money either, because the *haciendas* were idle. It was dangerous to go to the lowlands: one could be shot in error by soldiers or rebels. Some men dared to go to the *haciendas* by night to look for grain and sugar in the abandoned warehouses. Everyone was glad when in 1914 the fighting seemed to come to a halt.

On control and anomie

Four hundred years of regional society cannot be adequately accounted for in a few pages; nor is it possible, in a chapter with such a historical range, to avoid a degree of superficiality. Yet, in order to prevent excessive generality, the focus has been centred upon events and situations which determined land use and labour organisation in the Morelos Highlands.

The concepts *social force* and *power domain* were used with the purpose of setting regional society in the context of the changing wider society. Regional history could not be told, let alone explained, without taking into account external forces and overall domains, as well as their evolution and ruptures. The forces generated by the interests of the colonial government disappeared after independence. This had vital consequences. The local incumbents of authority roles had represented not mere individual ambitions but the policy of a colonial government which exerted an effective and persistent control, and which used its power to maintain a state of dynamic equilibrium among other social forces. All other categories and interest groups needed the support of the colonial government to pursue their goals and to achieve effectiveness in their respective power domains. In contrast, the nineteenth-century authorities (especially before Díaz's period) had a weak and ambiguous link with a national government which changed its personnel, policies and ideologies every two or three years. Furthermore, the local authorities no longer had any powerful supralocal institution—such as the Church or the army—to ally with and counterbalance the social forces responding to particularistic interests. Undermined by internal divisions and external political hostility, the Church lost its important role in education, *rites de passage*, finances and agriculture. As for the army, it had become a springboard for individual ambition rather than an instrument of national consolidation.

The power of the Church, and even more the power of the local authorities, had been solidly based on their control over an important part of the environment of the Indian population through the allocation of communal land and housing, the regulation of labour for *haciendas*, the disposal of surpluses for tribute and tithes, the orthodoxy of religious praxis and ideals. All this ended when tribute was abolished and the Indians could sell their labour in a free market, when the Inquisition was banned and religious freedom proclaimed. The *barrio* system could no longer be effective as a controlling device when all legal discrimination against ethnic categories—including that affecting residential patterns—was declared void, and even less when disentailment laws abolished communal property. While communal lands still existed, municipal councils could define their boundaries and extent, but this was not done according to a well defined general policy, only according to the interests and preferences of the councillors—and their personal alliances, which were frequently with landowners and planters. After the liberal victory the use of all lands became oriented towards the goal of accelerated capitalist growth. In turn, the absolute priority of this goal resulted in a *de facto* subordination of regional and national political power to the interests of national and foreign economic elites.

There is a sort of fatality in the sequence of Morelos's nineteenth-century history: municipal reforms, disentailment, land sales, prosperity of sugar enterprises, violence. At the turn of the century there were only two significant types of social actors operating in the Highland region: the land-owning entrepreneurs and the landless villagers. However, landowners lacked a peaceful mechanism to deal collectively with the villagers. The new controlling devices—such as wages, sharecropping contracts, or punishment—applied to individuals, not to groups; in addition, these devices were notoriously inadequate, in so far as plantations, *haciendas* and *ranchos* were incapable of absorbing all available labourers the whole year round. There were no roles of mediation: *alcaldes*, *mayordomos*, *gobernadores* had become redundant except in ceremonies. And *fiestas* had slowly lost their importance as an occasion for the symbolic convergence of heterogeneous interests. *Fiestas* were only for the 'Indians'—and this term was more than ever derogatory; their celebration often deepened the cleavages between classes.

The deterioration of the land situation, the critical problem of poverty and underemployment, the *anomie* in the relationship between the landed elite and the villagers finally implied that the latter could take no collective action *vis-à-vis* the planters and their clique except in terms of open conflict. It is no wonder that a fundamental strategy of post-revolutionary governments has been to develop collective mechanisms of manipulation and control in the rural areas: land reform devices, systems of municipal authority, schools, and vertically managed economic schemes. But the anomic burden of the past is still heavy on the shoulders of the Morelos Highlanders.[60]

Notes

1. For instance: Tlayacapan (nose of the earth), Atlatlahucan (between two ravines), Totolapan (turkeys on water), Tlalnepantla (place of terraces). See Robelo (1887).

2. On ancient Morelos history, see Müller (1949), Holt-Buttner (1962), Sanders and Price (1968), Martínez-Marín (1968), Fábregas (1969), Sotelo-Inclán (1970), Sáenz (1975), Carrasco (1976a).

3. There has never been a proper archaeological survey in the Highlands. Yet it is possible to locate stone walls in the hills by sight.

4. The *Codex Mendoza* describes the mechanics and geography of tribute payment in the Aztec confederacy (quoted by Sotelo-Inclán, 1970: 19 ff.; see also Riley, 1973: 97–9).

5. *Mercedes*: land grants awarded by the King of Spain.

6. *Encomienda* literally means *commission*: by this institution, the Crown put a set of Indians of one or more villages under the supervision of a Spaniard, so that he could take care of their proper Christianisation and moral welfare in exchange for their periodical labour services and tributes (Zavala and Miranda, 1954: 52 ff.).

7. See Furtado (1970: ch. II). The problem of capital accumulation in New Spain is discussed by Palerm (1975 and 1976).

8. The strip of land immediately below the Highlands (Cuautla–Amilpas–Yautepec–Jojutla) had at least twelve sugar refineries at the beginning of the seventeenth century (Lira and Muro, 1976: 109–10). The other important sugar area was Veracruz, which was principally connected to overseas markets. The main market for Morelos sugar was Mexico City (Barrett and Schwartz, 1975: 536–7).

9. The marquisate ended exactly where the region of the Morelos Highlands—as defined in Chapter I—begins (García-Martínez, 1969: 31 ff.; Sotelo-Inclán, 1970: 43 ff., 49).

10. Some names: Diego de Ordaz, Antonio Serrano de Carmona, Christóbal de Cassasano (the founder of a sugar mill which four centuries later still bears his name), Pedro Sánchez Farfán, Leonardo Lomelín (see Zavala and Castelo, 1939–45, vol. III, no. XCV; Martínez-Marín, 1968: 30; Riley, 1973: 24 ff.; Hernández-Orive, 1973: 93 ff.).

11. Other parts of Morelos had an earlier evangelisation from Franciscan and Dominican friars, who founded separate ecclesiastical provinces; thus Augustinian influence was still another factor—additional to ecology and marquisate boundaries—in keeping the Highlands a distinct region.

12. Oral tradition maintains that the library which existed in Tlayacapan in the sixteenth century was superb, and that the big canvas with the image of Our Lady of Guadalupe still kept in Totolapan is the oldest reproduction of the Tepeyac virgin in the whole country. In addition, a sculpture of Christ, on the cross, said to have been

covered with a paste containing real human skin, soon acquired a miraculous reputation which transcended regional boundaries.

13. Rights of property for individual Spaniards had multifarious origins in New Spain (see Florescano, 1971: 30 ff.; Gerhard, 1975: 572).

14. The size of most private estates in the Highlands did not seem to have exceeded one *caballería (41 hectares)* of agricultural land or one *estancia de ganado menor* (776 hectares) of pasture land. (*Mercedes* were given in these terms.) See A.G.N. *Mercedes.* 1542, vol. 1, exp. 408, f. 190v; *op. cit.*, 1542, vol. 1, exp. 395, f. 183v; *op. cit.*, 1542, vol. 1, exp. 396, f. 184v; *op. cit.*, 1567, vol. 9, f. 239v.–240; *op. cit.*, 1581, vol. 11, f. 32v; *op. cit.*, 1592, vol. 18, f. 90; *op. cit.*,1607, vol. 25, f. 419v; *op. cit.*, 1641, vol. 42, f. 22; *Tierras*, 1797–1800, vol. 1626, exp. 1, f. 35.

15. The town of Atlatlahucan, which became a municipal seat after independence, was a satellite of Tlayacapan during the colonial period (see A.G.N., *Indios*, 1591, vol. 6, part 2, exp. 198, f. 45). The fortress-like aspect of its monastery seems to indicate that in the early colonial period it was used as a frontier outpost.

16. Land titles for several Highland communities are registered in A.G.N., *Tierras*, vol. 1553.

17. During the decades that immediately followed the conquest, the Central-Mexico Indian nobility kept much of their own private land; but soon after they sold it or else were forcefully expropiated (Florescano, 1971: 65–6).

18. See A.G.N., *Tierras*, 1746–7, vol. 1472, exp. 1, f. 18. This process of 'independence' was not unique: in Morelos, the number of heads (*cabeceras*) grew from thirty-three in 1646 to ninety-four in 1800 (Gerhard, 1975: 575). Yet Tlayacapan did not only become autonomous, but absorbed settlements which previously were more important than itself; in addition, it received the title of *villa*, which was very rarely given to Indian settlements.

19. The local rituals for ancestors prescribe the provision of food for them at certain dates, which has to be put in specially designed dishes (see Chapter VII).

20. There is mention of Tlayacapan 'Indian merchants' in the late sixteenth century (see A.G.N., *Indios*, 1592, vol. 6, part 2, exp. 458, f. 102).

21. A.G.N., *Indios*, 1583, vol. 2, exp. 979, f. 225v.

22. Tlalnepantla declined during the seventeenth century, but became an important settlement in the late colonial period. In 1777 there were 604 resident families in that town, of which 547 were Indian and fifty-seven Spanish and *mestizo* (half-caste). See A.I.N.A.H., Papeles de Francisco del Paso y Troncoso, Legajo 100, pp. 337–41, 1777). Today, the same town has only two-thirds of its eighteenth-century population. The late colonial importance of Tlalnepantla is shown by the fact that the title of *ciudad* (city) was given to it.

23. The model thus does not refer to the surviving groups of hunters and collectors, nor to dispersed settlements of people practising slash-and-burn cultivation. And, given that a crucial variable of the model is the existence of communal land, obviously it does not apply to non-Indian villages or towns.

24. For detailed accounts of the constellation of variables constituting the model of the corporate community, see Wolf (1955, 1957, 1965); also Aguirre-Beltrán (1952), Nash (1967), Wolf and Hansen (1972). The term was coined by Wolf; other authors refer to 'closed', 'centripetal', 'cohesive' villages.

25. It seems plausible to think that the interpretation of *fiestas* as mechanisms of

redistribution was inspired by the classical analysis of *potlatch* on the North Pacific coast. It is therefore interesting to note that these analyses have been substantially challenged. For instance, Piddocke (1968) has argued that *potlatch* occasions responded to the need to exchange a variety of scarce resources, which due to ecological reasons were unequally produced by different participant groups, or not produced at all by some of them.

26. In Peru, Viceroy Toledo promulgated a detailed legislation for organising 'corporate' Indian communities (Toledo, 1924). In rural Europe, communal organisation also seems to have been originally used as a mechanism of State power (Weber, 1974: 78 ff.).

27. Tributes to the Crown were paid by Highland villages in food (cereals). See Del Paso y Troncoso, 1939–42, vol. X: 57 ff.; Martínez-Marín, 1968: 32 ff.).

28. Aguirre-Beltrán (1953: 37) contends that the substitution of 'hispanized' offices for the *tlatoque* was a rather slow process. Zavala and Miranda (1954: 64) point out that the native chiefs were replaced by elected municipal councils in the middle of the sixteenth century; but also add that the *tlatoani* sometimes co-existed with the new system of indirect rule and performed some special tasks in the administration (*ibid.*: 60–1). Carrasco (1961) has argued that the decay of the indigenous nobility and the rise of commoners to civil offices occurred early in the colonial period as a result both of the Spanish efforts to reduce the whole Indian population to a levelled peasant status, and of the commoners' interests in eliminating the restrictions imposed upon them. Recently, Gibson (1967: 166, 168–9, 194) has provided evidence showing that, at least in central Mexico, the *tlatoque* lost their importance before the end of the sixteenth century. The abolition of rights to personal tributes for the *tlatoque* was of central importance in this process (*ibid.*: 186). See also Carrasco (1957: 13 ff.). A different situation occurred in the region of Oaxaca, where many Indian noblemen kept their wealth and authority until the nineteenth century (Taylor, 1972).

29. For details on relationships between Indian officers and the Church, see Simpson (1952: 70–1), Aguirre-Beltrán (1953: 41) Gibson, 1967: 117–18, 181–2, 186, 218). On the Morelos Highlands, see A.G.N., *Indios*, 1591, vol. 5, exp. 1042, f. 337–337v.

30. There were many types of *cofradías*—including those organised by occupational groups in larger towns, those which congregated land-owners and rural labourers, etc. Here I am concerned with *barrio cofradías*.

31. The very existence of *calpultin* in pre-conquest Mexico has been questioned by contemporary ethno-historians (Brigitte Boehm and José Lameiras: personal communication).

32. This seems to indicate that *barrios* were colonial mechanisms of articulation between complementary areas. In certain villages of Texcoco, *barrios* only came into existence when the population was under the control and supervision of Spaniards (Angel Palerm: personal communication). The importance of *barrios* in the town of Totolapan decreased when its communal property was invaded by *hacienda* San Diego Huixtla.

33. A.G.N., *Tierras*, 1969, vol. 1591, exp. 4, f. 10. It is interesting that the names which appear in this document are still found in *Tlayacapan* (Vida, Santa María, Nopaltitla).

34. A property of the *fiesta* as an event is that not all the social forces operating in

the context are necessarily relevant to the behaviour and its patterning in the event itself. The analysis of how certain forces are relevant in a situation or event is perforce crude when one is handling historical data.

35. Relevant documents are kept in the Archivo Municipal de Tlayacapan.

36. See the following documents at the Archivo General de la Nacion: *Mercedes*, 1544, vol. 2, exp. 609, f. 242; *Indios*, 1592, vol. 6, part 1, exp. 130, f. 32; *Indios*, 1591, vol. 6, part 2, exp. 190, f. 44; *Indios*, 1592, vol. 6, part 1, exp. 309, f. 83v; *Congregaciones*, exp. 262, f. 126, 9 Dec. 1604; *Tierras*, 1746–7, vol. 1472, exp. 1, f. 18; *Tierras*, 1797–1800, vol. 1626, exp. 1, f. 35; *Tierras*, 1796, vol. 2550, exp. 9, f. 10; *Tierras*, 1801–7, vol. 1694, exp. 1, f. 430. See also Zavala and Castelo, 1965, vol. 7, nos. CCXC and CDXXXIII; Chavez-Orozco, 1953: 149. In addition, oral tradition testifies that there was a permanent feud, which lasted two centuries, between Highlands villages and their eastern neighbours of the region Chalco-Amecameca, which started when the latter diverted a river which previously reached the land surrounding Atlatlahucan.

37. See A.S.R.A., *Atlatlahucan*, 1921.

38. Matamoros, a local curate, became the regional head of the movement.

39. For an early case in the Highlands, see A.G.N., *Indios*, 1590, vol. 4, exp. 498, f. 149v.

40. '*. . . por ser bastante ladinos en el idioma castellano' (A.G.N., Tierras*, 1806–8, vol. 1625, exp. 5, f. 14). The word *ladino* is applied to 'westernised' people in some regions of Middle America.

41. Bourbon reforms had suppressed municipalities; the Cádiz constitution restored them, and the first Mexican constitution (1824) maintained the system (Cunniff, 1966: 59–61).

42. This document is kept in the files of the town hall of Tlayacapan. I am grateful to my colleague Enrique Valencia for drawing my attention to it.

43. In addition, there were no bishops who could ordain new priests (Mecham, 1966: 346). Later, in the middle of the century, most priests lived in urban places (González, 1956c: 363 ff.).

44. It seems likely that these lay specialists developed from the catechists used by the missionaries for instruction purposes.

45. Pertinent figures on land expropriation can be found in Whetten (1948: 86–9), Silva Herzog (1964: 122–3), Huizer (1970: 12).

46. According to Palerm (1975), the main cause behind the growth of export agriculture was the collapse of the world silver market: it was necessary to substitute silver production for other sources of foreign money.

47. For instance, the owner of the *haciendas* of Treinta and Coahuixtla was Manuel Araoz, a politician of high standing in Portfirio Díaz's circle. The *haciendas* of Tenextepango, San Carlos, Cocoyoc and Pantitlán were the property of the De la Torre family, one of whose members married Portfirio Díaz's daughter.

48. See the first chapter of his novel *El Zarco*. On the general process of village engulfment and destruction, see Womack (1969: 44–5); also Rascón and Cordera (1972: *passim*). Warman (1976: 53 ff.) gives a description of *hacienda* administration, technology and life-styles.

49. According to the National Agrarian Commission (1925), Coatepec was formed by communal land taken from Ahuatlán, Nepopoalco and San Nicolás. In 1895, its

bought by Vicente Escamilla, who remained as the owner during the Revolution (see A.S.R.A., Nepopoalco, 1895, 1925).

50. *Haciendas*: Mayorazgo, Buenavista, San Carlos, Oacalco, Coatepec, Tepantongo/Diego Huixtla (the main territory of the first four lay outside the Morelos Highland region); *ranchos*: La Mohonera, Santa Cruz, Alotepec, Mancera, Jagüeycillo, Del Moral, Del Astillero, Zacapulas, Los Manzanos, El Jagüey, El Jaral, Actipa, La Capula.

51. La Cascada (San Miguel Tlatetelco, in the *municipio* of Atlatlahucan) was the main station in the Highlands.

52. The land of Pantitlán was subsequently absorbed by the San Carlos *hacienda*.

53. The ensuing description of the Morelos revolution is mostly based on Womack's classic study of the Zapata movement. Other useful sources are Sotelo-Inclán (1970: chs. XXV–XXVI), Huizer (1970: 14–36), Warman (1976: 104 ff.).

54. Escandón substantially decreased the already benevolent taxes on plantations, and increased them for small farmers and traders (Womack, 1969: 53).

55. Madero was a member of a rich family of northern Mexico, who contended against Díaz in the 1900 national presidential election. Díaz was re-elected by fraud, at which Madero summoned all democratic forces of Mexico to overthrow the old dictator.

56. This account of Rosendo's life history was given to me by his son, Angel Rojas Santamaria.

57. A steep hill south-west of Tlayacapan—where Cortés met local resistance.

58. A *real* was equivalent to the eighth part of a *peso*: 12½ Mexican cents.

59. Compare Rosendo's work as *tariero* with Warman's description of *aparceria* in eastern Morelos (1976: 80).

60. Polarisation of contradictory interests is not a sufficient cause for the outbreak of social violence. Nor—on the other hand—can sheer repression explain social order. The existence of a minimal set of accepted social norms, or the lack of it, has also to be taken into account for the analysis of persisting peace or strife. Hence, the concept of *anomie* becomes relevant for the understanding of the situation which preceded the Morelos revolution, although I am in disagreement with Durkheim's equation of stable equilibrium to 'normality'.

IV
THE PROMISED LAND

La tierra se reparte con un rifle. [Pablo Neruda,1970:126]

Winds of agrarian change

The 'constitutionalist'[1] movement which formally took over the government in 1914 was made up of a heterogeneous lot. Venustiano Carranza was an old politician—senator and governor under Díaz's regime—heading groups of middle-class reformers and potential entrepreneurs who openly disliked Zapata and were in turn despised by him. Pancho Villa was a mass leader whose followers had no real ideology but were sympathetic to Zapatism and its agrarian tenets. Finally, there was a set of radicals: Lucio Blanco, Francisco Mügica, Miguel Mendoza López, who proposed more or less coherent programmes of radical change in city and countryside—but who did not form an organised group. All these groups met at a national revolutionary conference (the Convención de Aguascalientes). Zapata did not send any official representatives to the conference, though he agreed that some of his men could attend.

To his surprise, the conference officially adopted 'the principles of the *Plan de Ayala*' as a part of its political programme. Then a government was elected, and Manuel Palafox, one of Zapata's key men, was appointed Minister of Agriculture. Carranza refused to recognise this government, and his troops withdrew from the capital. Civil war was again imminent, but Palafox decided to proceed as if the situation were perfectly normal. He decreed a national agrarian reform which had as its main purpose to give the communties their land back. In fact this reform was put into effect only in the zones controlled by Zapata: Morelos,

Guerrero, Puebla, parts of the state of Mexico and Oaxaca.

The years 1914–15 are remembered by Morelos villagers as a period of victory and hope. The only brigade which came from Mexico City was a civil one, formed by students of the Faculty of Agriculture, who came as volunteers to mark the boundaries of the recovered communal property. Even though the work of boundary setting did not satisfy everyone—most village representatives disagreed about limits of neighbouring lands—the vast majority of families now had access to a cultivable plot. In the Highlands the *ranchos* and *haciendas* of the vicinity were taken over—thus Atlatlahucaners occupied the *ranchos* El Jaral and Alotepec, and Tlayacapaners did the same with *hacienda* San Carlos. Zapata attempted to create a sort of decentralised democracy in which local people took collective decisions. There were of course some opportunists who managed to get more land by doing Zapata's officers favours. In Tlayacapan, people say that some of the men who nowadays boast of having been *zapatistas* were really procurers and sycophants. 'Real warriors died in battle; survivors were not interested in revolutionary dreams.' But most people were content; they could grow maize and beans and pulses again—and in larger quantities than before. The harvest of 1915 was plentiful: there was food for everybody in Morelos. Nobody forced villagers to export. Zapata wanted to re-open the sugar mills as a source of public wealth, but the people did not want to grow sugar cane.

The end of Zapata. If the Convención government and Zapata appeared to the nation as agrarian heroes, Carranza had to present a similar image. His rationale for withdrawing from the Convención was that he could not come to terms with bandits; yet he produced a Land Reform Decree in January 1915 under which (*a*) alienation of village land was nullified; (*b*) villages having the political status of independent settlements (*pueblos*, *villas* and *ciudades*) could be granted land from large adjacent properties, if it was needed for the villagers' welfare. This decree attempted both to appease the rural poor and to reach a compromise with the *haciendas*. On the one hand, it excluded many settlements from land distribution; for instance, land could not be claimed by 'company towns' within plantations or by hamlets formed by dwellings of permanent *hacienda* workers. On the other hand, the decree included an important innovation: villages could claim land not only as a restoration of their old communal properties (*restitución*) but also as a grant (*dotación*). This circumvented the numerous difficulties inherent in land restitution: frequently old communal deeds were difficult or impossible to find, and/or boundaries impossible to define taking as a basis colonial measurements and references. In addition, the *dotación* procedures implied that large land holdings could be expropriated even if they had not invaded communal land. However, expropriation was to be limited to 'adjacent' properties, and landowners had the possibility of injunction—which could delay and even reverse the process (Whetten, 1948: 114–15). Another difference between Carranza's law and the Convención land reform was that the latter allowed the villagers to decide whether they wanted to work land individually or in common, whereas Carranza's made it clear that land should be divided into individual plots (Womack, 1969: 229).

The Carranza decree was not put into practice very eagerly in those parts of the country under the control of his army. It was, rather, a pretext for Carranza's generals to expropriate latifundia for their own personal profit (Whetten, 1948: 116). Yet it established a precedent from which there was no return, even after Zapata's forces lost ground.

The Convención government did not last long. Carranza's troops re-entered Mexico City in October 1915. Carranza continued his agrarian propaganda: a National Agrarian Commission was set up, and promises of justice were loudly proclaimed. In addition, a formidable army was sent against Morelos. Zapata's army went back to guerrilla resistance. The Highlands became an area of strategic importance. As in the time of the Spanish conquest, the Morelos people resisted the invaders, perched on hills and steep rocks, using all sorts of missiles. On 16 July 1916 a lengthy battle took place in Tlayacapan—on the same spot that local tradition holds the battle against Cortés occurred.

Carranza's men—now called the *nacionales*—entered Morelos in 1916; during 1917 Zapata recovered ground and regained most parts of the state; in 1918 the *nacionales* came back with new strength. The Morelos revolutionaries, already short of supplies, were decimated by a wave of Spanish flu—which also took the lives of many villagers. The climate was again one of hunger and fear. Besides, the Carranza government had issued a new Republican Constitution, which included radical articles of land reform. The 1915 decree still stood, although its application was slow. Many rebels defected, or quietly laid down their arms. The *nacional* generals turned into agricultural entrepreneurs, sowed sugar cane and prepared the mills for work. Even the old planters started coming back. In 1919 Zapata was murdered in an ambush.

A new deal. Many revolutionary radicals who joined the armies of Villa or Carranza had become alienated from *zapatismo* because of its localism and stubbornness. It was accepted that the land problem was a pressing one, but the nation had to solve a great many other problems and achieve unity. They rejected *zapatismo*, but kept defending its basic principles—which were partly reflected in the 1915 decree and later in the 1917 constitution. In its twenty-seventh article the constitution declared all land and all water to be national property, which could be transmitted by the nation to private individuals, but always subject to expropriation in the interest of public welfare. It added that the formula for the solution of the agrarian problem should include the restoration of land and the outright grant of land to the villages, as well as the enactment of federal State laws to break up latifundia.

So far so good. Yet the Carranza government did little to distribute any land. For the people of the Morelos Highlands it was clear that the killers of Zapata could not be revolutionaries. The myth of Zapata's resurrection started: it was said that he had appeared riding a white horse across the Ajusco. Some people asserted that he was hiding somewhere—in Arabia, perhaps, for the newspapers were full of stories about rebellions and guerrillas in the Middle East. Rosendo Rojas was among the people who read newspapers aloud to his fellow villagers; but he had little faith in the survival of Zapata. The plantations slowly started production, and Rosendo took occasional jobs as a *tariero*. Work was still rather scarce. Some mills or *trapiches* bought small amounts of sugar cane to produce *piloncillo* (brown sugar) and alcohol. But other commercial crops flourished in the lowlands: bananas, for instance. Rosendo took jobs with banana growers which consisted of planting and cultivating a thousand banana trees. It was called *enmilarse*: to take one thousand (*mil*). He became a famous *enmilado* and managed several contracts of this type simultaneously.

The 'new deal' of constitutionalist land reform became a reality when, in 1920, Alvaro Obregón, the general who had defeated Villa's troops in the north, replaced Carranza after murdering him. Obregón was a living synthesis of the contradictions of the Mexican revolution. He was a middle-class rancher who wanted modern capitalism to flourish in the country as strongly as in the United States; yet he advocated social change and used a language as radical as contemporary Soviet leaders. For him, it was clear that the building of a national consensus could not be attained without the support of the rural population. When Obregón mobilised the majority of the army against Carranza, Gildardo

Magaña—Zapata's successor—reorganised the dispersed Morelos forces and joined him. After Obregón's victory the Morelos guerrillas were recognised as 'the Division of the South'. Then Soto y Gama—one of the *zapatista* ideologies—organised the National Agrarista Party, with the open protection of Obregón. The party won several seats in Parliament and pushed through a number of agrarian decrees. One of them established the National Agrarian Commission, as well as commissions in every state, in charge of attending to petitions for land distribution. Other decrees restated the rights of all independent settlements to apply for land, as both *restitución* and *dotación*. Private property was not abolished, but holdings could be expropriated if necessary in the public interest. In addition, Dr Parres, who had acted as 'Zapata's leading medical attendant' (Womack, 1969: 367), was appointed Governor of Morelos. He also issued several state decrees advocating agrarian reform. Agents of the National Agrarista Party visited the villages encouraging people to organise committees and apply for land.

Slowly the *zapatista* army disbanded. The purpose of the fight had been to get land—and apparently the land was theirs. But the legal procedures, the *tempo* of land distribution, the way in which it was implemented, showed that the most important actors in the new agrarian process were outside the villages. Moreover, the fact that the process had to start all over again implied that Zapata's distribution had been illegal (Warman, 1976: 151–2). The 1917 constitution made it clear that the owner of all natural resources was the nation. And national representatives would be responsible for deciding about the use of resources. Fifty years later, many villagers still think that there is a contradiction somewhere. 'They gave us the land—the land they wanted, not the land we wanted. And they still have the legal means to take it away from us.' Perhaps it is because of this perception that many peasants have kept their old guns at hand.

Land distribution in the Highlands[2]

The formal steps in the process of land distribution were as follows:

1. Once a village had decided to claim land, a Comité Particular Administrativo of six members was elected in a public assembly. This committee was to undertake all the requisite procedures in the name of the whole village.

2. The Comité Particular made a formal petition to the Comisión Local Agraria. This was another committee which functioned at the state level (Morelos, in this case) in matters of land distribution.

3. The Comisión Local Agraria asked the state governor to certify that the village petitioning land was an effective separate settlement, and not a *barrio* or section of a town or village.

4. The governor, after consulting state files, produced an answer.

5. The Local Agrarian Commission sent a visitor (usually an agricultural engineer, or at least a qualified topographer) to survey the land, take a census of the village in order to distinguish landowners from landless people, and ascertain the general social and economic situation—including the features of any large land holdings in the vicinity of the village. If the petition was of *restitución*, the Local Agrarian Commission studied the historical documents provided by the villagers and the National Archives concerning communal property.

6. The visitor produced a detailed report and included a series of recommendations for action.

7. The Comisión Local Agraria sent both the report and the recommendations to the governor as well as its own conclusions. In addition, the owners of large land holdings which the visitor thought should be expropriated (if any) were informed of that possibility.

8. A period ensued to allow landowners to make a case against expropriation of their land, if they had any legal objection. (They invariably had.)

9. A resolution was dictated by the governor (*Resolución gubernamental*) which took into account the report, recommendations, and legal objections.

10. If the governor had resolved that some land should be given to the villagers, the latter could take provisional possession of it, but the land could not yet be formally divided into plots.

11. The complete file of proceedings was sent to the Comisión Nacional Agraria—the committee in charge of land distribution at the national level.

12. The Comisión Nacional Agraria conducted any investigations it thought pertinent to the case.

13. The Comisión Nacional Agraria produced a report and a formal statement to be submitted to the President of the Republic.

14. The President gave a final resolution (*Resolución presidencial*).

15. Land was finally given to villagers (if that was what had been decided) and cultivable land was divided into family plots.[3]

The implementation of these fifteen steps took different forms in different regions and specific villages. To give an idea of local variations I present a fairly detailed account of land reform in three Highland settlements: the town of Atlatlahucan, the town of Tlayacapan, and the small village of Asuncion Ahuatlan.

The agrarian reform in Atlatlahucan

On 24 January 1920, a group of neighbours wrote to the Governor of the State of Morelos demanding that proceedings for restitution of their land should be started. Apparently the Governor never responded to this letter. Several months later the Ayudante Municipal sent another petition, this time to the Local Agrarian Commission in Cuernavaca. The Ayudante declared that documentary evidence existed of their rights over the land surrounding the town, as well as over the land of two *haciendas*, San Diego Huixtla and Tepantongo. These two *haciendas* had in 1780 been sold to the Indians of Atlatlahucan by the archbishopric of Mexico, but later, in 1856, the tenant in charge of the *haciendas* acquired property rights under the protection of disentailment laws. The Ayudante also claimed that the town of Atlatlahucan was the lawful owner of several water sources. These streams had their headwaters in the Sierra of Amecameca, in the east of the Morelos Highlands, and, according to oral tradition, their courses had been illegally diverted towards other regions.

This time the state government reacted to the petition, and on 21 October 1920 the Local Agrarian Commission (henceforth L.A.C.) appointed an expert (*ingeniero*) in charge of the technical aspects of the proceedings. Yet the actual agrarian reform procedures did not start until January 1921, when a committee of neighbours was formally elected. The committee sent the Governor a detailed report of all the tracts of land claimed by the town. A quick census revealed that there were 1,298 inhabitants in Atlatlahucan, grouped into 306 homesteads; of which, sixty-seven families had no land at all.With the *ingeniero's* help the committee worked out the boundaries of existing plots and presented a report to the L.A.C. The report showed that there were among the neighbours a few 'big shots', holding plots of fifty hectares or more (17 June 1921). Later, on 4 July, the *ingeniero* reported to the L.A.C. that there were at least 365 men (including heads of families and adult males) who should be granted land; for, even if most of them were holding plots, the size of the plots was not enough to provide a decent living.

But the proceedings soon ran into difficulties. On 20 July 1921 the L.A.C. asked Maurilio Fabre, the owner of San Diego Huixtla and Tepantongo to present the deeds of the estates. Fabre sent all his documents two weeks later, vigorously stating his case against Atlatlahucan. His argument, many times presented to the commission (in 1921, 1922 and 1923), was that he had bought the *hacienda* from Francisco Flores, who in turn had acquired the property directly from the archbishopric. The Atlatlahucan Indians had intended to get the land; but, Fabre said, they never did because they lacked the money to purchase it.

Meanwhile, on 28 July 1921 the L.A.C. sent a new *ingeniero*, who contradicted all previous reports and declared that the only two possible sources of land for the town were the *hacienda* of Casasano and the *rancho* El Jaral, both to the south. He stated that other tracts of land should be given to other *pueblos* in the region. The L.A.C. sent a letter to the National Agrarian Commission (N.A.C.) on 6 August asking for advice and declaring that the land demanded by Atlatlahucan was simultaneously under claim by other settlements. This confusion was furthered by the inefficiency of town committees. A new *ingeniero*, Milton Rubio, sent by the N.A.C., wrote a new list of the tracts of land claimed by the town: the southern *ranchos* of El Jaral and Santa Cruz (owner, Eduardo Idrac); the north-eastern *rancho* of Alotepec (owner, Gabriel Sanvicente); the north-eastern *haciendas* of San Diego Huixtla and Tepantongo. The list was sent to the L.A.C. on 15 September, but the Ayudante and the committee also sent a list, which added to Rubio's account more than twenty *ranchos* and sizable holdings. A week later Rubio informed the L.A.C. that disagreements over boundaries and land rights in Atlatlahucan were due to the rivalry of two local *caciques*. He reported that apparently other villages (Tetelcingo, Nepantla and Yecapixtla) had really invaded Atlatlahucan property, and that the situation was serious for the 359 citizens who had a right to be granted land.

At this point, groups of Atlatlahucan people, organised by the aforementioned *caciques*, started invading the land of other *pueblos*, Oaxtepec, Yecapixtla, Pazulco, Tepalcingo. This caused a storm of protest. In their turn the citizens of Atlatahucan complained that all the land surveys had been carried out by racketeers, and they reiterated their petition for restitution of their land (letters to the L.A.C., November and December 1921). Yet the procedures made no progress: according to the L.A.C., the original Atlatlahucan communal land titles (confirmed by the Viceroy in 1539) did not cover as much territory as the Atlatahucan people claimed. On the other hand, Sanvicente, the owner of Rancho Alotepec, did not have his deeds in order; and the matter of San Diego Huixtla-Tepantongo, in the L.A.C.'s opinion, could probably be resolved in favour of Atlatlahucan. In any case, things seemed to indicate that a *dotación*, not a *restitución*, would take place.

And so it did. On 4 January 1922 Rubio produced a report with the following points. There were 365 people who should be granted land: 305 family heads plus sixty unmarried adult males. The soil around Atlatlahucan was not very good: stony hills and seasonal land, with no hope of irrigation. Each grantee should be given ten hectares of arable ground plus some forest land, for which he recommended the expropriation of some tracts of Rancho La Mohonera (100 hectares), Rancho Santa Cruz (125 hectares), Rancho El Jaral (71.5 hectares) and the *hacienda* San Diego Huixtla-Tepantongo (100 hectares). These tracts, together with the small plots already owned by local people, would be sufficient for the present. The general level of poverty was critical: the town was suffering from the scourge of usurers and market monopolists. The main local products were maize, beans, chillies, tomatoes, pumpkins and some water melons, which were sold to Cuautla through La Cascada railway station.

The L.A.C. notified all this to the Governor on 7 April, and he gave his *resolución* three days later. The Governor was more generous than the engineer: he conceded 792·415 hectares to the villagers, of which 548·5376 should be seasonal cultivable land, and the rest hill and pasture. The land holdings affected were San Carlos (461.36 hectares of seasonal land), El Jaral (84.17 of seasonal land), Alotepec (46.87 of hill country), San Diego Huixtla (200 of hill and pastures). The L.A.C. was authorised to mark boundaries and start the *ditación provisional.*

Yet the business slowed down again. The invasion of other villages' land continued, and there were small outbursts of violence. On the other hand, the Atlatlahucan committee did not accept that land distribution was made in terms of *dotación*. They claimed that both El Jaral and San Diego Huixtla were theirs. A document of 1716 was produced according to which the Indian community of Oaxtepec had sold El Jaral to Atlatlahucan for 50 *pesos*. In turn the owners of El Jaral replied that they had the contract by which the mayor of Tlayacapan had sold the ranch to their ancestors; but the Atlatlahucaners denied the validity of it. In addition to this apparently insoluble problem the L.A.C. had to attend to the protests of both Sanvicente (the owner of Alotepec) and Fabre (San Diego's). Sanvicente alleged that the Governor's resolution was illegal, because his ranch (the main part of it) was situated in the state of Mexico. He managed to rally on his side the L.A.C. of this state, and also the mayor of a neighbouring Mexico *municipio*, Tepetlixpa. Fabre demanded protection (*amparo*) 4
from a federal court, and got it, so procedures had to start all over again as far as his property was concerned.

The rest of 1922, and the whole of 1923 and 1924, were spent in a multiplicity of disputes. Angry letters went from neighbours to both the L.A.C. and the N.A.C., complaining about the corruption and inefficiency of their surveyors; from the Morelos L.A.C. to its Mexico state colleagues, pointing out that land distribution could in fact transcend state boundaries; from several landowners to all sorts of authorities, accusing villagers of invasions; from surrounding villages, accusing Atlatlahucan of the same thing. Even the Secretaria de Gobernación (Ministry of Interior) intervened to disapprove of invasions. The issue of the Amecameca rivers was again brought up by the Atlatlahucan Administrative Committee and the matter was sent to the authorities of the state of Mexico, who never replied.

In January 1925 the National Agrarian Commission took things into its own hands and spent two more years trying to sort them out, not very successfully. There was confusion about (*a*) which land had really belonged to Atlatlahucan, (*b*) which invasions (from people of their town and others from near by *pueblos*) should be 'tolerated', since the land was to be given to the people anyway, (*c*) which large land holding should be expropriated, and for which villages. Confusion was in part created by the Atlatlahucaners themselves, who kept changing their specific petitions. It was hinted by several neighbours that one of the local bosses was fostering confusion to delay procedures and personally profiting from the existing ambiguity—for instance, by monopolising the marketing of produce in invaded land, and by himself using plots of undefined property, like ancient Church property. Finally President Calles issued a *resolución presidencial*, on 27 April 1927, declaring a *dotación* of 1,847 hectares, taking them from San Diego Huixtla-Tepantongo (700), Coahuixtla—*ranchos* Mancera and Jagüeycillo (229), Casasano (212), Xalpa (42), Santa Cruz (175), El Jaral (27) and San Carlos (462). Out of these, 943.8 hectares were seasonal agricultural land, 903.2 forest, hill and pasture. The *resolución* pointed out that the problem of water would be met by a regional plan.

The *resolución* did not finish all the troubles. Although the *ejido* authorities had already been appointed and set to work by August, disputes over boundaries continued, both between Atlatlahucan and other villages, and among Atlatlahucaners themselves. A source of trouble was the fact that people who had obtained good land in the *dotación provisional* did not want it properly measured and marked lest their holdings were reduced (in fact, many did have to be reduced). The completion of boundary-setting in agricultural plots took seven years, and still a great many *ejidatarios* were dissatisfied. The boundaries of non-agricultural land never were set very clearly.

Agrarian reform in Tlayacapan

The petition for land restitution was made by the citizens of Tlayacapan on 1 January 1921. A year passed before the Governor certified, on 2 August 1922, that Tlayacapan was an autonomous *villa*. The villagers were not remarkable for their patience: they proceeded to invade the lands of Oacalco *hacienda*, just as some years before they had taken land from Pantitlán—San Carlos. The argument of the Tlayacapan people was that all this land had previously been communal property. The Oacalco manager threatened the invaders, but did not dare to take direct action against them—the days of *hacienda* impunity seemed to be over. Instead the *hacienda* reported them to the state authorities and demanded that rent should be paid by the squatters, without getting any reply. Six months after the encroachment, the L.A.C. sent a surveyor, who measured 633 hectares (practically all the occupied terrain) and authorised the villagers to work in it provisionally—but making it clear that an investigation would be held. Nothing was said about rent. A proper investigation was not started until five years later.

In 1928 the Oacalco *hacienda*, slowly recovering from economic troubles in the recent past, and now with aggressive new owners,[5] returned to the fight. Its property had been severely reduced because of land grants to lowland towns (Yautepec, Itzamatitlan), so one of the partners, Luis Creel, was commissioned to start a lawsuit and recover whatever he could. At his insistent demand the L.A.C. appointed a visitor to Tlayacapan, who reported (on 14 February 1928) that 1,096 of the townspeople were already in possession of 3,913 hectares (in the form of both communal and private property), of which only 327 were cultivable, and that agriculture was the only source of livelihood, apart from the raising of a few cows and sheep. He suggested the distribution of at least 473 hectares of cultivable irrigable land among 331 adult males, taking for this three large fields: Cacahuatlán (from the *hacienda* Pantitlán-San Carlos), and El Chirimoyo and Chapultepec (from (Oacalco).

There followed a year of legal strife between Creel and the villagers. Creel argued, firstly, that Tlayacapan's communal land was more than enough for its needs. Secondly, that Tlayacapan was more than seven kilometres from Oacalco, so the latter was legally immune from expropriation to the benefit of the town. Thirdly, that the Oacalco sugar-cane fields and refinery were again in operation and provided an important source of income for farmers and industrial workers—which could be jeopardised if the spoliation of land and water continued. Fourthly, that the Oacalco refinery produced sugar and alcohol—two key products in the Morelos economy—and provided an important source of income for the State itself, as it paid taxes of more than 71,000 *pesos* per year.

The Tlayacapan neighbours answered by simply restating their rights over the land: they produced a copy of the sixteenth-century title deeds, which turned out to be genuine. This convinced the L.A.C. and the Governor, who issued his *resolución* on 8 March 1929. It confirmed communal rights over 1,371 hectares already held by the town (and not at present claimed by anyone else), and also over the Cacahuatlán field. In addition, other tracts of land from Oacalco and San Carlos were to be granted (as *dotación*) to landless citizens, who were declared to be 211 in number. The *resolución presidencial* was not long in following: it came out on 22 August 1929, leaving the Governor's decision practically unchanged. Thus 211 people received the Cacahuatlán field (296 hectares) plus 263 hectares from Oacalco and another 72 from San Carlos-Pantitlán—a total of 631 hectares, practically the same land that had been originally taken.

Land reform in Asunción Ahuatlán

The Comité Particular Administrativo was elected on 29 November 1922, and the same day a petition was sent to the L.A.C., suggesting that land be taken from the Coatepec *hacienda*. The L.A.C. wrote to the owner asking for plans and relevant documents on 20 December. The Governor confirmed on 15 January 1923 that Ahuatlán was an independent *pueblo*. A month later the L.A.C. visitor had produced a first report, stating that the population numbered eighty-six, of whom forty (twenty-five family heads plus fifteen unmarried adult males) should be given an *ejido* plot. He added that there was an acute problem between Ahuatlán and Totolapan—because, in the immediate past, Ahuatlán had been considered a *barrio* of that town but was never included in its plans for land reform. In a second report the visitor described the characteristics of Coatepec land: it had previously been mountainous forest, but widespread felling of trees had converted it into a desolate spot. The soil was shallow, calcareous and rough, with patches of *tepetate*. It could be used as third-class seasonal agricultural land but, productivity being low under such conditions, the minimum size of plots should be from ten to fifteen hectares. That meant that at least 400 hectares should be exported from Coatepec. It was also necessary to provide the villagers with communal ownership over the local water deposit (*jagüey*) which at the time was in the private hands of a rancher, who asked a fee for the right of taking water. Many people could not afford to pay. Landless people who managed to hire out their labour got only 0.75 pesos per day—the same as it cost to hire a donkey. Some people used the land surrounding the village (which had been part of the *fundo legal*, space earmarked for housing in colonial times) for maize cultivation—but the soil was also shallow, and it was in total less than fifty hectares.

The L.A.C. sent its conclusions to the Governor on 6 July 1923, and his *resolución provisional* came out on the 12th. The amount of land was reduced to 240 hectares, and the numbers of grantees was not specified. On 30 July the villagers provisionally took possession of the land.

There the matter rested for two years, until the Ahuatlán people took it up again: they wanted the water problem solved (letter to L.A.C., 31 July 1925) and also more land (15 October 1925). There followed two more years of silence, until the N.A.C. sent a visitor who produced a new report (9 June 1927). The number of people had decreased in those five years to seventy-two (twenty-seven family heads and ten unmarried adult males). Only 25 per cent of the land taken from the *hacienda* was cultivable; the rest was shrub and meagre pasture. On the other hand, Coatepec still had 1,438 hectares. As for the problem of water, a solution had become urgent.

The ruling of the President of the Republic was given on 1 September 1927. It reduced the grant to 230 hectares—for thirty-seven grantees—of which 57.5 were agricultural and 172.5 pasture and shrub. The neighbours took definite possession on 2 December.

The overall characteristics of the process. The cases provide a basis for discussion of the general pattern of the process, supplemented by data from other Highland communities.

1. It was a long, complicated process. On average, from the day of petition to the day of the President's resolution, it took almost six years (De la Peña, 1980: Table 4.1). The reasons for this slowness were many. All *pueblos* asked for a restitución: but this demanded the presentation of documents which were both authentic and clear. The authenticity could not always be proved. The documents presented by San Andrés, for instance, turned out to be false—including Viceroy Velasco's signature. Other villages lost precious months looking for documents in the Archivo General de la Nación. On the other hand, it was difficult to mark precise boundaries on the basis of colonial documents. References to 'a big tree', 'a stone in the shape of a cow', 'a plot known as St Cuthbert', were difficult to locate after 400 years. In addition, there was no consensus among neighbouring villages about boundaries. Things were further complicated by the fact that, under Zapata's regime in 1914–15, plots had been occupied by people of one village which were now claimed by people from another village. In any case, agrarian officers tended to be suspicious of traditional boundaries. Land reform should be sanctioned not by tradition but by the new State's authority. Finally, delay was also caused by pressure on the part of landowners, and by quarrels among villagers themselves—this last point is underlined by frequent changes in the composition of the village committees.

2. The government did not always respond to the land petitions very quickly. Governor Parres usually produced a resolution in two or three years. Later, it was President Calles (1924–28), Obregón's successor, who had to sign the final resolution. But Calles was not enthusiastic about land distribution, and procedures were delayed on purpose. After Calles, only two settlements required a presidential resolution: Tlayacapan, in 1929, under Portes Gil's mandate—it went through quickly, as Portes Gil was in favour of reform; San Nicolás, in 1936, under Lázaro Cárdenas's presidency, which also went through quickly, for the same reason. As for San Miguel el Fuerte, which applied in 1933—under President Abelardo Rodríguez, who did not sympathise with the agrarian cause—it did not get even a provisional *dotación.* It

became more obvious that land reform was not a local process but a movement directed and controlled from above.

3. The amount of cultivable land granted per head was very small—3.2 hectares on average over 1925–36 (De la Peña, 1980: Table 4.2). It was not until 1934, under Cárdenas, that a detailed Agrarian Code was implemented. The code stated that the minimum amount per head would be eight hectares. However, Morelos had practically completed the process of land reform before the code was even conceived.

4. The landowners opposed land distribution in all cases. Maurilio Fabre (owner of San Diego Huixtla-Tepantongo) claimed that the agrarian censuses taken in Tlatetelco, Tepetlixpita and Totolapan were inaccurate. Angela Sánchez (El Astillero) kept delaying the process of *dotación* to Tlatetelco by accusing the villagers of forceful invasion. The Escamilla family (Coatepec) demanded documented proofs of the *hacienda* invasion of Nepopoalco land, and later demanded that the census of the village be checked. Oacalco repeated its argument of being an industrial unit of public utility. It also used the strategy of denying the validity of the San Sebastián and San Agustín censuses, and argued that these villages were not independent settlements but *barrios* of Tlayacapan. In many cases the landowners' tactics seem to have been partially effective: the National Agrarian Commission usually reduced the amount of land awarded to the people by the governor's resolution. The two reviving local sugar *haciendas*, Oacalco and San Carlos, seem to have been protected at the level of the National Commission. Several letters from officers of the commission recommended the expropriation of other holdings instead of taking land from the sugar *haciendas*. And the Tlayacapan process was blocked for eight years—until President Portes Gil's breakthrough in 1929.

5. The problem of water scarcity remained unsolved. Practically all settlements explicitly demanded water facilities for both drinking and irrigation. The commission made promises of wells and dams which were never fulfilled.

6. The process was practically stopped after 1930. Two villages—San Miguel el Fuerte and San Nicolás el Vigía—applied for grants in 1933 and 1934. The case of San Miguel was not even researched until 1943. In 1947 a governor's resolution declared that twenty-seven families of San Miguel had 'the right' to land grants—but that there were no neighbouring holdings which could be expropriated. San Nicolás had better luck: President Cárdenas's resolution came in

less than two years, and each grantee was given eight hectares of seasonal land. (This in spite of the manoeuvres of Coatepec *hacienda*.) On the other hand, eight villages (Tepetlixpita, Atlatlahucan, Tlatetelco, Asunción Ahuatlán, Totolapan, San José de los Laureles, Tlayacapan and San Agustín Amatlipac) applied for an *ampliación* (expansion) between 1934 and 1937. According to the Agrarian Code, a village had the right of *ampliación* when the growth of population required more land. Yet only the towns of Totolapan and Tlayacapan got a positive answer. Other cases were blocked by *haciendas* (for instance, Mariano Riva-Palacio, the owner of Buenavista, successfully opposed Tepetlixpita's expansion) or were simply told that they had the right to land but that none was available in the vicinity. (This meant that people could get land outside the region, but nobody wanted to travel a thousand kilometres and take the risk of being granted a piece of jungle or desert.) The expansion of Totolapan implied the creation of a new settlement, Nicolás Zapata,[6] in 1939, for the land expropriated from Mayorazgo *hacienda* was rather distant for people to travel every day from town. (Mayorazgo managed to deliver only 507 hectares, sixty less than decreed.) The expansion of Tlayacapan was petitioned in 1936, and the governor resolved favourably in 1937; yet the Oacalco *hacienda*—from which land was to be taken—managed to delay the President's resolution until 1941 and to decrease the size of the *ampliación* from 127 to 73 hectares.[7]

7. The process implied continuous conflict of different types. Conflicts between villagers and landowners were natural—though the holdings invaded by landless people were sometimes not big *ranchos* or *haciendas* but small farms. It was also common for people from different villages to quarrel over boundaries; practically all settlements were involved in quarrels of this type.

Another kind of conflict resulted from villagers allying with big landowners against other villagers. For instance the *rancho* El Moral had on its side some Totolapan people who worked for it, against other people from the same town who were asking for the seizure of El Moral land. Similarly, the sharecroppers of Buenavista defended the *hacienda* against villagers of Tepetlixpita. The *rancho* El Jaral was able to manipulate the rivalry between sharecroppers of Atlatlahucan and San Juan Texcalpan to delay expropriation.

Sometimes these conflicts resulted in open violence. When San Nicolás was granted its land in 1936 the sharecroppers of the *haciendas* Coatepec and Mayorazgo refused to leave. Further, they demanded of the government that the land for San Nicolás should be taken from other tracts of the *hacienda*. The government agreed. But time went by and the new boundaries were not set. In 1938, taking advantage of the confusion, a group of people from Juchitepec (a town in the state of Mexico) came and invaded the zone. A part of the area invaded corresponded to lands of Tlatetelco and Totolapan, whose boundaries were also unclear. In revenge, several Totolapaners invaded what was left of *hacienda* Buenavista. Then fighting among all these groups broke out. Several men were killed. In 1941 the National Agrarian Commission sent a topographer who had to flee for his life. He went back with 200 soldiers. The Juchitepec people had to leave, in spite of the letters that the C.T.M. (Confederacion de Trabajadores Mexicanos, a workers' confederation which is a sector of the ruling Revolutionary Party) sent on their behalf.[8] Land could be finally measured and bounded. But personal hostility remains to this day.

The intervention of the C.T.M. was not exceptional. Other sectors of government, or of the Revolutionary Party—the 'official party', which has control over practically all government employees in Mexico—were present in many disputes. For instance, in a dispute over boundaries among Totolapan, Asunción Ahuatlán and Nicolás Zapata, which began in the 1940s and lasted until the '50s, the C.N.C. (Confederación Nacional Campesina, the peasant sector of the party) took sides with Totolapan. On the other hand, the 'Federation of Revolutionary Workers and Peasants of Mexico' sent a letter to the Agrarian Commission in favour of Asunción Ahuatlán.[9] Even in minor disputes among neighbours (over land or for other reasons) the intervention of government or party officials is not infrequent. Moreover, the legal grounds may in some cases be less important than the direct support of powerful groups or individuals. Again, as in colonial times, access to land and resources depends upon alliances of local people with outside forces and their representatives.

The organisation of collective property

The National Census uses the generic word *ejido* to name all land in Morelos which falls outside the regime of private property. The colonial *ejidos* (from the Latin *exitus*) were those tracts immediately outside the village, which were used in common (not necessarily for cultivation). Later the word *ejido* was colloquially applied to all types of communal land. According to the last census, the Morelos Highlands *municipios* have 11,263 hectares of *ejido* (Atlatlahucan, 2,303; Tlalnepantla, 1,486.3; Tlayacapan, 2,417.5; Totolapan, 5,505.8), of which 40 per cent is agricultural land. These figures should include both the land granted and/or restored by the government after the revolution and that which was in fact in the possession of the communities at the time of the distribution and was recognised as being theirs. (The land distributed by decree between 1925 and 1936 amounted to 7,390.7 hectares—see De la Peña 1980: Table 4.2; 580 more were granted as *ampliaciónes*.) However, the figures are probably underestimated.

No registered collective property can be legally alienated, rented, or used directly as a commodity. An explicit decree of the president of the Republic is needed to change this situation in a particular case.

Agricultural and non-agricultural *ejido* lands are administered differently. Non-agricultural land is held by the community as a whole. Every person resident in a settlement has the right to free use of the pasture land of the community, or the firewood collected on communal land. An exception, however, is made for some tracts in the northern *municipios* of Tlalnepantla and Totolapan, where pine trees grow. The felling of this timber is controlled by federal inspectors. Tracts of barren land on the fringes of settlements are used as football fields, fairgrounds and for other recreational activities.

In contrast, agricultural *ejido* land has been divided into plots to be separately cultivated by individual holders. At present, practically none of these plots is vacant. Most of them were occupied immediately after the Agrarian Reform was implemented in the region. The basis for plot allocation was that the recipient should not own any land; in fact, many small proprietors managed to obtain *ejido* holdings through manipulation of friendship and kinship ties with the local authorities. Many sons of private landholders were also allocated *ejido* plots, and later inherited their parents' private property. All this was facilitated by the virtual non-existence of land records at the time.

The size of the distributed plots varied from place to place. In many small settlements the cultivable *ejido* land was divided equally among all the applicants. (For instance, in San José de los Laureles all irrigated *ejido* plots have 2.5 hectares.) In other places early applicants got comparatively bigger plots. But everywhere, as a general rule, *ejido* holders (*ejidatarios*) control only small plots which do not exceed two or three hectares (see Table 4.1).

Rights over a plot of *ejido* land can be inherited but not sold or otherwise alienated. In addition, the integrity of the plot has to be

Table 4.1
Distribution of cultivable *ejido* land in two towns (*in hectares*)

	Number of people			
Size of plots	*Tlayacapan*	*Totolapan*	*Sub-total*	*%*
Less than 1	197	48	245	46.2
1–2	112	62	174	32.8
2–3	19	49	68	12.8
3–4	2	22	24	4.7
4–5	0	12	12	2.2
5–7	0	7	7	1.3
Total	330	200	530	100.0

Source: Local *ejido* files (1971)

maintained; only one heir is legally permitted. The land cannot be left idle. If there is no heir, or the holder abandons it, the plot is allocated to another person.

All these measures tend to prevent both the emergence of exceedingly small holdings—which would result from dividing the plots—and the monopoly of plots. But, in fact, many *ejido* farmers divide the land among all their sons, even if only one of them is the formal heir for legal purposes. [10]Rental of *ejido* holdings is also frequent.

Each village or town has a local committee (*comisariado ejidal*) elected from among the local holders every three years which is in charge of re-allocating abandoned plots, defending the *ejidatarios'* rights against squatters, and arbitrating in disputes over boundaries. The *comisariado ejidal* is jurally both a representative of the local *ejido* holders and an executive officer of the federal land reform authorities. (The latter have throughout the years changed their name from National Agrarian Commission to Department of Agrarian Affairs and Colonisation to Ministry of Agrarian Reform (Secretaría de la Reforma Agraria).

In some villages there is also a separate committee in charge of that part of the *ejido* which supposedly belonged to the community in colonial times. In these villages, people differentiate between these types of fields, which are called *comunal* (communal), and the rest of the *ejido*, which is called by this name. The allocation of cultivable communal land is sometimes altogether controlled by local authorities, and is given to people for short periods of time. In Tlayacapan, for instance, there are

103 registered *comuneros*, who renew their permits every year. San José de los Laureles possesses 400 agricultural hectares distributed in a similar manner. On the other hand, the sons of *ejidatarios* are sometimes given a small piece of communal land to build a house on.

Tierra comunal—particularly non-agricultural land—was not always carefully measured and marked by the federal authorities. Further, in some communities it was never formally confirmed or registered. This has caused conflicts, some of which are still unsettled. Communal land in San Juan Texcalpan was confirmed as such, but boundaries were never set—which has resulted in continuous quarrels with Atlatlahucan. San José de los Laureles has a standing dispute with Tlayacapan for a hill (a valuable hill: it contains a gravel pit), and another one with Tlalnepantla for forest land. The town of Tlalnepantla claims rights over about 6,000 hectares which overlap with land claimed by Tlayacapan, Tepoztlán, Milpa Alta and Juchitepec.[11] The feud with Milpa Alta is particularly important: it involves water springs and valuable timber forests. On the other hand, some plots within the communal tracts are claimed by individuals as private property. The slowness of the federal authorities to solve the problem conclusively (the Tlalnepantla affair has been going on for forty years) may be due to bureaucratic inefficiency, but in fact it has been functional for centralised power as a means of maintaining and increasing the dependency of local people. Periodically the Tlalnepantla Comité de Bienes Comunales (which is normally elected every three years) visit the Mexico City and Cuernavaca authorities; they are told that their case is progressing; letters are shown to them in which different officers discuss aspects of the problem. Once they saw the draft of a President's resolution in their favour—which never came out. From time to time an officer visits the region, takes a census, interviews local authorities and produces a report. (There have been at least four such visits.) Nothing happens. If there are signs of violence, people are threatened with the army; in fact the army passes by from time to time.

The organisation of private property

The triumphant revolution by no means attempted to destroy private property in the countryside, simply to put limits to it. On the other hand, the return to a pre-capitalist mode of production was not a goal of the new government. Even if some of the ideologies which inspired a number of revolutionary leaders included the notion of an idealised, primitive agrarian community,[12] the tendency prevailing at the end was one of economic modernisation combined with policies of social welfare. *Haciendas* were attacked because they had produced social instability among the landless peasants; because sometimes—this was not the case in Morelos—they were not very efficient economically, and because their owners were clearly Díaz's partisans and operated in the context of an alliance between the government of Díaz, the interests of American and European imperialism, and the national upper classes.[13] The revolution was now to create a new, efficient, nationalistic bourgeoisie. As far as the countryside was concerned, the new policies aimed at producing a large number of small farmers who would have access to plots of reasonable size and to a national market where they could sell their produce. From this perspective, the inalienability of the *ejidos* was conceived of as a barrier against appropriation of land and the formation of big estates.

But 'moderate' private plots were completely approved of. In this section, I shall refer to the lot of the lowland private estates, and to the situation of post-revolutionary private property in the Highlands.

The suffering and survival of the haciendas. The planters and their successors were back in Morelos as soon as Zapata was dead. Many Highlanders who had taken land from San Carlos-Pantitlán in 1914–15 took fright and left the plots; yet the *hacienda* invited them to go back and grow maize as *medieros* (sharecroppers)—i.e. in exchange for half the harvest if ploughs, oxen and seed were provided by the *hacienda*. The plots no longer received irrigation, though San Carlos still controlled the sources and the channel system which brought water from the Yautepec river. Obviously the *hacienda* was waiting for the occasion to go back to sugar cane and recover its old power. Even when land reform was announced by Obregón throughout the country it was not dismayed. The restored power of San Carlos seemed so well established that the villagers of San Andrés refused to take *ejido* land expropriated from this *hacienda* lest they were hurt or even killed.

However, by 1927 San Carlos's property had been subject to nine expropriatory decrees (three definitive and six provisional). These expropriations amounted to 1,853 hectares; but the *hacienda* still held 1,737 hectares, of which 693 were irrigated.[14]

As for *hacienda* Oacalco, in 1925 it was holding 3,779 hectares, of which 1,105 were under cultivation (sugar cane and maize). This *hacienda* also commanded several water springs, yielding more than 500 litres per second—i.e. enough for irrigating about 400 hectares, though in fact only 100 were irrigated. Water was used exclusively for sugar cane. Oacalco was slowly going back to the pre-revolution pattern of organisation. Both sugar cane and maize were produced by *gañanes* and *tarieros* who came from the surrounding lowland and Highland villages and received a daily wage which fluctuated between one and two *pesos*. It also used sharecroppers to grow maize in the less productive tracts of seasonal land (i.e. that which produced sixty hectolitres for each hectolitre sown; the best seasonal land produced eighty to one).[15] Oacalco possessed seventy-nine mules and thirty-six oxen; also two tractors and eighteen modern steel ploughs. The company owning the *hacienda* ('Caja de Préstamos para Obras de Irrigación y Fomento de la Agricultura') was prepared to resume a leading productive role in Morelos. The same company also held land in other parts of the state (in Santa Rosa Cocoyotla and San Vicente Chiconcuac). As had been the case with San Carlos, Oacalco also suffered several provisional expropriations between 1925 and 1928, but still managed to keep more than 2,000 hectares.[16]

In the late 1920s one could see a pattern developing in Morelos which seemed to favour the revival of sugar-cane *haciendas*. In contrast, those *haciendas* which were not directly employed in sugar production lost practically all their land in the same period. For instance, 2,822.9 hectares had been expropriated from San Diego Huixtla between 1923 and 1928, which left this Highland *hacienda* with only 324.1 hectares of hill and second-class seasonal land.[17] Coatepec fared similarly. Buenavista managed to cut down practically all valuable trees in the vicinity of Totolapan before its end.

Wage labour and poverty. For the Highlanders this pattern of 'agrarian reform' meant that their economic situation was not substantially better than before the revolution. They received rather bad unproductive land. Even those *ejidatarios* who received plots in the lowlands were not really better off, because their plots were not irrigated, since the control of the water was still in the hands of the *hacienda*. Furthermore, some of the plots were malaria swamps, and the villagers had to work at draining and levelling before they could use them. Thus many decided to leave the region and look for jobs elsewhere. Others stayed and scratched a living from their plots and wage labour.

Productivity in the Highlands was lower than in the valley: less than twenty hectolitres of maize for one sown. Wages ranged between 0.50

and 0.75 *pesos* per day. A day's wages would buy one litre of maize (the price for one hectolitre being seventy *pesos*) or half a litre of black beans: just enough for a family to survive on if the father could find work four or five days every week. But wage labour in the Highlands was scarce, since there were very few people controlling sizable plots of land. In fact there were even fewer medium-sized landowners than twenty years before. Thus the flow of labourers to the lowlands again became inevitable. Sugar-cane production was already creating a demand, and wages were higher. In addition, Highlanders continued carrying out some of their traditional tasks: the cultivation of fruit trees and pulses in domestic orchards and small terraces, collecting wood from the hills, raising small herds of cows and sheep in the hills and hens and turkeys in the back yards.

A new type of latifundia. In 1929 President Portes Gil granted land to several Morelos communities—both in the mountains and in the valley—and this resulted in further expropriation of the territories of *haciendas* San Carlos and Oacalco. President Cárdenas did the same in 1934 to 1937. San Carlos disappeared as an agricultural enterprise soon after—during the years of the world-wide slump—as the owners sold most of the land which had been left to them; but Oacalco has persisted until the present day. In 1936 Oacalco held 411 hectares, most of which were used in sugar-cane fields and its corresponding irrigation system.[18] In 1941 Tlayacapan got an *ampliación* of seventy-three hectares, taken from Oacalco. However, the *hacienda* managed to keep control of the use of the land, as well as a lot of the other surrounding fields, because it still controlled the supply of water. Villagers had to cultivate sugar cane and sell the produce to the Oacalco mill, lest irrigation be cut off from their fields. The government accepted this because Oacalco (and other *haciendas* in similar circumstances, such as Casasano) alleged that the high costs of re-making channels and ditches, and installing electric pumps, had to be recovered. Thus the Oacalco mill could grind about 60,000 tons of sugar cane in 1937, of which only 10,000 were produced in the *hacienda's* fields. The rest was produced by villagers of five lowland (Yautepec, Santa Inés Oacalco, Los Arcos, Itzamatitlan, Ignacio Bastida) and two Highland (Tlayacapan and San José de los Laureles) communities.[19]

A new type of latifundia began to emerge, controlling labour and produce not through direct appropriation of land but by more subtle means. One was investment in irrigation systems. In 1953 Oacalco expanded its influence by drilling a well near the lowland part of the Tlayacapan *ejido* and building channels which would make it possible to irrigate the whole of it. The *ejidatarios* would pay for this by discounting a percentage of the value of sugar cane sold to the Oacalco mill. Twenty years later they are still paying this debt, without having ever managed to get a clear accounting statement from Oacalco. In fact the latter claims that the debt has increased because of the costs of channel maintenance, electricity for pumping, and mortgaging of pumps.

Another strategem of the mill for keeping control over labour and produce is related to systems of organising sugar-cane production, which will be dealt with in Chapter VI. And of course, an important factor which initially helped to re-create the dependence of Highlanders on sugar refineries was poverty and lack of work opportunities in the Highlands.

Private property in the Highlands. There are in the Highland *municipios* 1,580 registered owners of private property, which covers 7,837 hectares.[20] As we have seen, many private holdings were occupied by landless people during the revolution. Later the government surveyed the land and confirmed property titles over fields which were not claimed as communal land. Many plots which had belonged to the local rich before the revolutionary period changed hands, because very few rich people came back. However, many small landowners retained their land. Both types of private holders share one characteristic: commonly they do not possess properly legalised deeds. For practical purposes, the authorities recognise ownership on the part of anyone paying taxes for a private plot, provided that nobody else claims property rights to it. This fact has caused problems when the rights of inheritance are not clearly defined by the family of a deceased man—i.e. when there is no consensus on who is to keep the plot. Then he who pays the higher bribe to the Public Property Registry gets the land.

All private holdings can be divided and sold at the will of the owner, and there are in fact a great many divisions and sales. The dynamics of the use of agrarian soil in the last decades will be described and analysed in the following chapters. For the moment, it is important to note that (*a*) the overwhelming majority of plots (almost 80 per cent) are less than five hectares in extent; (*b*) statistics of land distribution do not reflect actual control, because renting has become a common practice; (*c*) there are a growing number of people who have no land, or a negligible amount of it.

Table 4.2
The Morelos Highlands: size of private property (in hectares)

	Less than 1	*1.1–5*	*5.1–10*	*10.1–25*	*25.1–50*	*50.1–100*	*100.1–200*	*Total*
Atlatlahucan								
No. of owners	51	187	83	49	13	2	3	388
Area	38.4	546.4	626.8	728.9	471.3	174	411	388
Tlalnepantla								
No. of owners	138	119	39	40	1	–	–	337
Area	81.2	295.9	318.6	616.5	36.9	–	–	1,349.1
Tlayacapan								
No. of owners	143	262	57	30	3	–	–	495
Area	79.3	646.3	408.7	433.8	104.1	–	–	1,672.4
Totolapan								
No. of owners	123	193	29	10	5	–	–	360
Area	78.3	443.3	223.8	172	141.7	–	288	1,348.1
Sub-total								
No. of owners	455	761	208	129	22	2	3	1,580
Area	277.2	1,932.1	1,577.9	1,951.2	612.3	174	1,312.3	7,837
Percentages								
No. of owners	28.79	48.16	13.16	8.16	1.39	0.12	0.18	99.95
Area	3.53	24.65	20.13	24.89	7.81	2.22	16.74	99.97

Source: 1970 agricultural census

There are only five holdings in the Highlands of more than 100 hectares. Four are mostly hills and pasture land; their owners live in Mexico City and pay little attention to them.[21] The other one is an estate near the town of Atlatlahucan, where the nephew of an ex-president of the republic has a country house, stables, a private well and poultry farms. Near Tlayacapan, the son of an ex-governor of the state of Morelos bought twenty-two hectares in 1960 to build weekend cottages and sell them to city dwellers. This was the beginning of a slow but steady process of land monopoly for tourist projects, some of which have been promoted by politicians.[22]

On revolutionary change

The revolutionary outbreak in Morelos was both triggered off and limited by a situation of contextual anomie: by a crisis of co-operation and coexistence between predatory entrepreneurs and villagers demanding land and employment. The emergent local movement lacked the organising capacity to deal with the established system. The temporary viability of *zapatismo* was due to the villagers' alliances with the discontented middle classes and, to a lesser extent, the emergent proletariat. Locally, these forces provided the leadership of *zapatismo*. Nationally, they provided the general conditions for a Morelos revolution. Zapata would have been crushed by the federal army had the latter not been fighting, first against Madero and later against the Constitutionalists. Without the Convención truce, Zapata's utopian land reform would probably never have taken place. And, finally, without the support of national forces, Zapata lost ground. After his death Morelos's history took a course which was the result of the renegotiation of local guerrillas with an array of social actors which included not only fellow revolutionaries but also representatives of the *ancien régime*.

Such renegotiation is patent in the process of land distribution. It was organised and controlled not by landless peasants but by government representatives. The government was interested in appeasing the peasantry by responding to their demands; but it established complicated legal procedures in order to keep control over each step taken by the villagers. Legal procedures were a symbol of government authority: it was the national revolutionary law, not the pressure of local *zapatismo*, which generated justice. Local committees could ask for justice, but, as far as direct action was concerned, all executive powers were—and still are—vested in extra-local officers. In addition, local committees—turned into *comisariados ejidales* after the *ejidos* were established—provided a device for the government to be present in the subsequent history of the land grantees. All *ejidatarios*—and the Comisariado checks this—automatically become members of the C.N.C. (Confederación Nacional Campesina), which is in turn a sector of the official party—the party which, from 1930 to the present, has won *all* elections in Morelos and 95 per cent of the elections in the rest of the country. When *ejidatarios* have had problems (for instance, in cases of *ampliación*, or renewed disputes over boundaries, or in looking for agricultural bank credits) it is convenient to have official support from the leaders of the C.N.C., or from other important figures in the party. The terms of the alliance between government and villagers seems to allocate quite a precise role to each party: for the villagers it is to ask the government representatives to grant 'favours', even if they only do what is part of their ordinary duties. The proverbial inefficiency of bureaucracy at all levels strengthens these roles: nothing seems to progress in a government office if one does not have a personal link with a functionary, who—as a 'favour'—makes things efficient for his friend or protégé. This is true not only in land matters, but also in welfare services and development works, such as education and health, roads, electricity, commercial schemes, all of which are organised by the government.

In the years after the revolution an important strategy of the new federal authorities was not to give local policital offices to ex-*zapatistas*. Moreover, some important offices were given to local people who fought against Zapata. Two examples follow.

Aniceto Carrera

Born in Tlayacapan, 1898. Illiterate. His father farmed a small terrace of his own property in addition to working in sugar *haciendas*. His mother was a domestic servant in a local rancher's house. When Aniceto was fifteen he was *enganchado* ('hooked') by the federal army. In 1913–14 he fought against revolutionaries in the south-western states of Mexico: Quintana Roo, Yucatan, Campeche, Tabasco. In 1915 he 'became a revolutionary', because his brigade joined the Constitutionalists. He was then sent to Morelos, and visited Tlayacapan and the Highlands several times, with anti-*zapatista* hunting parties. In 1917 he was demobilised and went to live in Tlayacapan, but soon had to flee because *zapatista* sympathisers wanted to kill him. He went as far as Atlixco (Puebla), where he rejoined the army. From 1917 to 1924 Aniceto led a highly mobile existence, following his brigade to practically every state in the republic. From 1924 to 1926 he was stationed in Mexico City. He was finally demobilised in 1926, and went to Tlayacapan, where—thanks to his superiors' recommendations—he obtained a post as a municipal police corporal. He kept this job until 1953. In 1954 he was appointed *juez de paz* (Justice of the Peace) and as such was consulted in the settling of petty disputes. After he went back to his home town, Aniceto played a minor role in practically all matters connecting Tlayacapan with the wider political system: agrarian distribution; campaigns of candidates for local, state and national offices; and the mobilisation of local labour for government works (mainly roads). His role has always been one of support for decisions which come from above through the official party. At present he leads a modest, if tranquil, existence, farming a four-hectare field which he inherited from his father and uncle.

Heladio Morales

Born in Tlayacapan, 1891. Went to school and learnt to read, write and count. Orphan since he was nine, started work at ten in a tinsmith's workshop. In 1912 he joined the federal army 'by accident': he had personal enemies who happened to be *zapatistas*, so Heladio ran to the other camp. In 1916 he went back to Tlayacapan and was appointed secretary of the municipal council by the *nacionales*. He kept the post until 1920, when the *zapatistas* were granted an amnesty; then he started a business supplying mangoes to Mexico City. In 1926 he was appointed municipal president, and in 1929 he was the secretary of the local committee in charge of the agrarian reform procedures. From 1926 to 1936 he was also the main organiser of the *fiesta* of the Lord of the Tres Caídas. His mango business continued until 1935, when he decided to set up a tinsmithy—the trade he had learned in his youth. The workshop prospered. At the same time Heladio farmed the land he got in the 1930s: an eleven-hectare private field (bought), and a 0.9 hectare irrigated *ejido* plot. In 1944 he founded a local association of sugar-cane producers and was its president until 1950. In that year he accepted a part-time job from the Oacalco *ingenio* as field inspector. In 1953 he resigned in order to devote more time to his own farming and to the tinsmithy.

A comparable case would be that of Ricardo del Valle, whose father had been Porfirio Díaz's tax collector for Totolapan. The Del Valle family left Morelos during the revolutionary period; in 1927 Ricardo went back to the Highlands and was given the same job his father had held. Later Ricardo was a local agrarian leader, a deputy in the state congress, and became one of the major landowners and merchants of the region.

Nowadays Aniceto, Heladio and Ricardo express sympathy for Zapata and his ideals, which they think the post-revolutionary governments have carried out in the best possible way. But local *zapatistas*–and also men like Rosendo Rojas, who did not take up arms but were supporters of Zapata—never wholly trusted people like them. In the last analysis, though, they had to accept them: Heladio and Aniceto represented the revolutionary government, and promoted land distribution. They participated in *fiestas* and were committed to local networks, even if their presence evoked the ghosts of Porfirian days.

In fact, Porfirian *hacendados* were present in the land distribution process not only as ghosts. They modified the process through injunction or bribery. In the end, they seemed defeated. The populist period of President Cárdenas (1934–40) dealt a mortal blow to the big Morelos latifundia in so far as most of their territory was given to villages. However, the new economic organisation brought about by the revolution still needed capitalist agricultural enterprises. These enterprises could develop activities which *ejidatarios* and petty private farmers could not possibly undertake, and would be protected—in an open way—by the government. In Morelos some *haciendas* were able to adapt themselves to the new situation through the adoption of new strategies. Other enterprises were altogether new. Both types of enterprise might be characterised by the term *neolatifundium*, which some writers have coined to stress the fact that these dynamic agrarian nuclei are not altogether dissimilar from the nineteenth-century landed institutions (see Stavenhagen *et al.*, 1968; Warman, 1975).[23] Modern enterprises do not have the political power *haciendas* did; but they still have a crucial influence on the use and organisation of village land and labour, thanks to their alliance with the modern Mexican State. Local agrarian leaders like Heladio Morales became employees of modern enterprises; Ricardo del Valle was both a government representative and a commercial entrepreneur: agrarian reform and capitalist entrepreneurship are not enemies, but bedfellows.

The Morelos Highlands in the 1930s

During the nineteenth century local commercial activities—which had been moderately important in the colonial period—gradually decayed. Muleteering became dangerous because of endless civil war and banditry. Finally, the railway to a large extent displaced other means of long-distance transport. In addition, the *haciendas* installed their own system (*tiendas de raya*), where labourers, by means of a coupon system, were encouraged to acquire food, alcoholic beverages, clothes, and incur heavy debts. These stores offered the best prices and were strong competition for most local petty merchants. Just before the revolution there were only a handful of established merchants in the Highlands—the 'rich' mentioned in Rosendo Rojas's biography—who were also moneylenders and small-scale cattle raisers, and usually lived in the municipal heads.

The 'rich' left during the revolution, and most of them never came back. Thus in 1930 social differentiation was slight. There was little money about. The region did not even boast a single medium-size store. Weekly open markets—an ancient tradition—continued to be held in the four main towns. The villagers sold cereals, green vegetables and fruits, and bought muslin, *manta* (a coarse cotton cloth) and sandals brought by small traders from Ozumba and Cuautla. Wooden ploughs and hoes also came from the outside (though some local craftsmen produced a small number of them), as well as draught and burden animals (donkeys, oxen and mules), which were sold by roving dealers. Clay dishes and ceremonial pottery were manufactured in Tlayacapan. *Adobes* and roof tiles for building houses were made by specialists in every community. Other local trades were bread-making, carpentry, etc. Every family collected firewood for cooking from the hills, and water from the *jagüey*; but each community also had someone who made his living by carrying water or by selling wood. To be a musician in a local band was also a reasonable source of income: *fiestas* needed music. The important local people were the priest—when there was one—the teacher, and the government officers.

Most people farmed small plots and grew maize, beans and chillies: the basic elements of the daily diet. (Meat was uncommon, except on holidays. It came from poultry and pigs kept in domestic back yards, as well as from local cattle.) Farming was not very profitable. For centuries people had supplemented their seasonal farming with dry-season wage labour in the irrigated lowlands. Now this activity was only partially available. Most of the land of sugar *haciendas* was being distributed, and irrigation had been suspended from most of it. In addition, several sugar *haciendas*—San Carlos was one of them—could not survive the critical capital scarcity and the market crisis brought about by the 1929 depression. *Haciendas* which were still active—such as Oacalco—had cut back and could not employ many people. In retaliation against the villagers' involvement with the Zapata movement and, later, agrarian reform, Oacalco—like other lowland *haciendas*—began to import labourers from other regions of Mexico—Guerrero and Oaxaca, mainly—and employed them rather than local people.

Such was the Highlands' social reality during the Cárdenas period. Appendix III reproduces Angel Rojas's (Rosendo's son) account of his own early years, which may provide an idea of local colour—and poverty. His outwardly happy view of childhood does not hide some dark aspects of contemporary Highland society. For instance, the poultry-stealing activities of Angel and his friends—as well as the cattle robberies by some of his fellows—have to be seen in the context of internal competition for scarce resources within and between villages. During 1930–40, the population, which had decreased during the revolutionary years, began to grow again. By 1940 most settlements had more or less the same number of people as in 1900 (Appendix II, Table 4). *Ejido* and private plots seemed insufficient to meet the demands of this growth. People with better land were envied by those less favoured. Disputes over boundaries arose constantly. Most villages asked for *ampliaciónes* but only two got a positive answer. As sons grew up they asked for 'a little corner' of their fathers', grandfathers' or uncles' plots. Bitter competition between brothers and cousins was frequent after 1940.

In 1936 the visitor sent by the Department of Agrarian Affairs (successor to the National Agrarian Commission) to report on the matter of the Tlayacapan *ampliación* was struck by the poverty of the district. His estimate was that a family needed, for mere subsistence, a yearly income of about 600 *pesos* (400 for food, 60 for clothing for three adults, 45 for clothing for three children, 100 for medicines and general expenses); and he reckoned that only a small minority in fact had this income.[24] It could be said that the populistic intentions of the Cárdenas period did not do much for a region like the Highlands. Cárdenas envisaged that the peasantry would produce for subsistence. But a peasantry like that of the Highlands could not survive without supplementary wage labour—unless perhaps there was more irrigation, or more land than was actually available.[25] It is not surprising that Zapata had planned to reorganise sugar plantations and mills.

Poverty had some paradoxical consequences. On the one hand, some patches of communal land on hills were opened for agricultural use and painfully cultivated by hoe (Ingham, 1968: 60). On the other, many people decided to go away and left their land idle. Some villages witnessed a massive exodus. For instance, in 1941 only thirteen out of sixty-three *ejido* grantees in San Andres Cuauhtempan were actually working their plots.[26] Yet many of these migrants—as we shall see—came back later.

The mention of bullrings held during fairs has a special importance in Angel Rojas's narrative. Further information on fairs will be given in Chapter VIII. These fairs and *fiestas* reveal the existence of wide interaction networks between settlements, which included fun, commercial exchange and also conflicts over *ejido* and communal land boundaries, over wandering cattle and over access to jobs in lowland fields. The amusing 'revolutionary' escapade of Angel and his young fellow villagers—their attempt to join Encarnación Barreto, a Morelos rebel—points to the precarious nature of social peace as late as 1942 or 1943 (it was around those years that the Mexican army conscripted youths for the Second World War). In fact Barreto's movement, locally known as *la bola chiquita* ('the little guerrilla'), found considerable support in eastern Morelos (Ramírez, 1973), and probably induced the government to intensify its strategies in order to reduce the inhabitants of Morelos to 'peacefulness and production' (the slogan it used during the 1940s).

As far as the Highlanders were concerned, they were willing to do anything—perhaps even to co-operate with the government—to get out of the critical situation in which they found themselves.

Notes

1. The movement adopted the enforcement of the National Constitution as an explicit banner.
2. The information upon which this section is based was obtained from the A.S.R.A., from decrees published in the *Diario Oficial del Estado de Morelos* (1921–30), and also from oral accounts given by witnesses of the process.
3. Division into plots was made formally obligatory by a decree of President Calles in 1925.
4. *Amparo*, legal recourse against abuse of executive authority.
5. After the revolution, Bonos Terrazas, an anonymous society representing the interests of an old landed family, bought Oacalco from Francisco Velez, the previous owner.
6. Emiliano Zapata's eldest son.
7. The people who got land in this *ampliación* founded the village of Pantitlan.
8. The Juchitepec people moved further north and founded a new settlement (Colonia 20 de Noviembre).
9. This was an interesting and complicated dispute. In 1935, Asunción Ahuatlán asked for an *ampliación*. The land for it should be taken from *hacienda* San José de Buenavista. In turn, the *hacienda* claimed that its land could no longer be expropriated, because what was left (150 hectares of mountain forest and ninety hectares of seasonal cultivable land) was within the limits of legal property. In 1938, Asunción insisted. They complained that Totolapan had applied for an *ampliación* after they did and had got it. A topographer visited the village and suggested twenty-three people to be granted land. The governor then ruled in favour of Asunción. Yet, at this point, the local authorities of Totolapan intervened saying that Asunción was not entitled to receive land because it was a *barrio* of their town; but they offered to share sixty hectares of their land as a proof of their goodwill and friendship. Asunción did not agree with this, but accepted the land—wild terrain which they opened up for cultivation. Seven years later, in 1946, Totolapan wanted its land back. The fighting

began. In 1947, the federal agrarian authorities ruled that Asunción Ahuatlán should keep the plots. But the villagers still insisted that a formal *ampliación* should be conceded to avoid further troubles.

10. A typical case: Emigdio Polanco died in 1971 in San José de los Laureles. Formally, his widow inherited his *ejido* plot, which has a size of 2.5 hectares. But he had made clear that the plot should be for practical purposes divided among his five sons. One of the sons had become a schoolteacher and lived outside the village, so he agreed that his part be divided amongst his four brothers, who then received 0.625 hectares each. Their mother lives with the youngest, but the other three also give some money to her from time to time.

11. The confusion is furthered by the fact that Tlayacapan land is partly registered as 'municipal property'—which makes it unclear whether it should be considered as collective inalienable property or not.

12. Apart from Zapata—who never was very clear about this point—perhaps the most notorious representative of this tendency was the journalist Ricardo Flores Magon, whose ideas presented a mixture of anarchism and anti-industrial socialism, a system he supposed had existed in the 'Indian' communities (see Flores Magón, 1970; esp. 35–9). Flores Magón's dictum 'Land and Liberty' (*Tierra y libertad*) was adopted as a slogan and battlecry by the *zapatistas*.

13. At the end of the nineteenth century, the mining industry of Mexico (including oil, copper, lead and tin) was in the hands of foreign companies which also dominated the production of rubber, sisal, and coffee, as well as the whole railway system.

14. A.S.R.A. *San Sebastián*, September 1927.

15. One hectolitre (100 litres of grain) of maize weighs about 0.073 tons. This is usually enough to sow one *yunta* (4.5 hectares),which in those days could be cultivated by a single man.

16. A.S.R.A. *San José de los Laureles*, July 1925.

17. A.S.R.A. *Atlatlahucan*, 1928.

18. A.S.R.A. *Tlayacapan*, 1936.

19. A.S.R.A. *Tlayacapan*, 1937.

20. The size of plots registered by the census more or less coincides with data taken from local sources.See Appendix II, Table 2.

21. These holdings are what is left of ***haciendas*** San Diego Huixtla, Coatepec, Mayorazgo and *rancho* Santa Cruz.

22. An ex-governor of Morelos has been repeatedly accused of seizing two fields (Campo Chiquito and Los Arcos) which are part of the Atlatlahucan *ejido* (see A.S.R.A. *Atlatlahucan*, 1963 and 1967).

23. In the 1950s, Emilio Alanís Patiño, a distinguished agricultural economist, already spoke of 'social latifundia' (quoted by Silva Herzog, 1964: 471).

24. A.S.R.A. *Tlayacapan*, Ampliación, 1936.

25. Ronfeldt's analysis of Puebla's sugar cane *ejidos* provides important points of comparison with the Morelos situation (1975).

26. A.S.R.A. *San Andrés Cuauhtempan*, 1941.

V
THE SOCIAL HERITAGE OF REVOLUTION

This chapter should be seen as the transition between the historical and contemporary parts of the book. Its purpose is to provide the basic information needed for the analytical use of the concepts *social force* and *power domain* in the period which began when land distribution ended.

Firstly, we examine the national strategies adopted by the post-revolutionary government *vis-á-vis* the agrarian actors, and the general impact of these strategies. Secondly, an account is given of the changes that occurred at the regional level concerning transport and external communication, penetration of the monetary economy and mass-produced goods, and opportunities for production and employment. Thirdly, the analysis focuses attention upon some crucial strategies adopted by regional actors: long-distance migration, occupational diversification, technological innovation, manipulation of kinship ties.

The term *actor* is used assuming that interacting individuals and groups have at least some possibilities of choice. The term *strategy* implies a decision-making process whereby decisions are revised and modified according to each actor's perception of the others' decisions (actual and possible), and according to their perceptions of regional and extra-regional power and economic conditions. If these perceptions are wrong, the strategy fails. In addition, choice is often limited by the consequences of the actors' previous decisions. For instance, demographic growth in the Highlands was in part the result of a strategy to increase the domestic labour force; yet the population boom had drastic repercussions on the ratio of people to land and therefore affected the patterns of land use. Similarly, the general adoption of modern agricultural technology reinforced money-oriented activities and social class differentiation.

The concept of *social class* cannot be used without referring to a macro-social framework. However, it cannot be assumed that the Morelos Highlands is a sort of microcosmic replication of Mexico; hence one should specify which are the regional connotations of the stratification process. Involvement with commercial agriculture has implied a tendency for land and labour to become commodities; yet direct land appropriation is formally limited by law. On the other hand, money for agricultural inputs is often obtained through heterogeneous non-agricultural activities, and local commercial agriculture presents a high level of risk and low level of capitalisation. These factors encourage the emergence of both undifferentiated (market-regulated) and differentiated social relationships, in so far as the latter are needed for gaining access to scarce resources—land, labour and multiple economic opportunities. This results in an intricate web of social alliances which—as we shall see in later chapters—prevent class polarisation. At the same time, the precarious conditions of production—which are in turn dependent on specific government policies—discourage accumulation of capital at the local level.

The manipulation of kinship links provides a crucial example of the manner in which differentiated social relationships develop. But there are many other types of differentiated links; for all of them the term *relationship of trust* will be used. Initially, trust could be defined as a persistent reciprocal expectation of benevolent behaviour among persons united by multi-stranded ('multiplex': serving more than one purpose) links. We shall come back to the term later.

Government strategies and post-revolutionary rural development

After the revolution, both state and municipal governments lost considerable strength as most of their attributes passed to the federation. Most taxes are now channelled into the federal treasury. In addition, the federal government is directly active at the local level through its Ministries and through all sorts of agencies—credit banks, rural warehouses, business enterprises, tourist development schemes. Even the enforcement of law and order at all levels is highly dependent on federal power through the action of the army and the federal police. As far as most rural municipalities are concerned their funds are hardly enough to pay meagre part-time salaries to the members of the municipal council and to one or two police constables. On the other hand, the official party—Partido Revolucionario Institucional (P.R.I.)—could in many ways be considered as an agency of the federal government. The party chooses candidates acceptable to it. In turn, the federal authorities control *ejidatarios* and unionised workers through their collective affiliation to the party.

Federal government strategies are neither totally consistent in themselves nor continuous over time. These inconsistencies respond to the multiplicity of interests which the Mexican State attempts to conciliate into a single social contract. The contradictory nature of these interests has in turn demanded the presence of a strong, indisputable referee—the federal government. Only a strong government can alternatively or simultaneously adopt strategies of a populist or capitalist nature. Thus, during the 1920s, the process of strengthening federal control was a national priority for those holding power. A particularly important strategy was that which responded to social demands—especially in the countryside—through the creation of centralised structures: an expanding educational system, and an agrarian reform apparatus. In addition to engineering centralisation, the top politicians of the period—Obregón and Calles—managed to be both popular leaders and clever businessmen. Their private agricultural enterprises expanded quickly—and their descendants are still notable tycoons. Calles had even begun to think—and say—that land distribution was after all not such a good idea. However, Calles lost all political influence after 1934, when President Lázaro Cárdenas took office.

Cárdenas, and many of his relatives, also became millionaires; but in his period (1934–40) the populist trend of the post-revolutionary Mexican State reached its peak. Cárdenas himself, almost twenty years later,[1] summarised his programme in four cardinal points: (*a*) agrarian reform, including good land for all peasants, and irrigation schemes for *ejidos*;[2] (*b*) the creation of a national market through government credit and marketing controls, and the organisation of co-operatives; (*c*) nationalisation of the main industries, including oil extraction; (*d*) free education for all, with an emphasis on technical schools and the training of teachers. Through education in particular, and government programmes in general, it was hoped that a new ideology would emerge and consolidate in the minds of the people which would be both 'socialist' (a term which was never very clearly defined) and nationalist.

The impetus of this programme fell off noticeably even before the end of Cárdenas's period. Formally, all subsequent governments have accepted these points (except the term 'socialism', which was soon expelled from the official vocabulary), but in practice they have not really been given top priority. Some (Silva Herzog, 1964; Hansen, 1971; Labastida, 1972) have seen official priorities from 1940 to 1970 as actually contradicting the Cárdenas programme, especially where the rural population is concerned. This change is often attributed to ideological differences between Cárdenas and his successors. But other factors should also be considered. In the early years Cárdenas's policies operated in the context of international economic crisis. It would have been useless to adopt a strategy of producing for the foreign market (as under Díaz's regime), because the foreign market was not buying. On the other hand, the home market was effectively reduced to the cities and largest towns, and even there the revolution had had devastating effects. This narrow market, in addition, was full of imported goods which did not seem easy to replace on a short-term programme. A wider home market had to be created; meanwhile the type of agricultural production which it was thought would be immediately favoured by the agrarian reform—i.e. one oriented towards local subsistence crops—appeared to be very convenient. On the other hand, scarcity of capital had severely diminished the rate of both urban and rural investment. Again, the convenience of *ejidos* and co-operatives functioning with low capital needs was patent. It should be added that Cárdenas's populism was a device for undermining foreign control of the country's economy and strengthening the hand of central government.

The overall situation was changed by the Second World War. International demand for Mexico's agricultural products was again high. Capital, especially United States capital, was available. Further, there was a chance to foster industrialisation through a policy of import substitution, as industrial products from Europe and the U.S.A. no longer flowed into the home market. The radical oratory of the Cárdenas government was replaced by slogans of 'National Unity'. Private investors, both Mexican and foreign, were reassured that private property was not in danger. Agrarian reform slowed down appreciably.[3] Government investment in rural areas (irrigation, credit banks, roads) actually increased, but was directed mainly to benefiting export-oriented—usually privately owned—holdings. For instance, from 1940 to 1960, 80 per cent of investment in irrigation and 50 per cent of that in transport works was allocated to areas in the north and west controlled by large landowners (Aguilar Mora and Medina, 1972).

Economic growth from 1940 to 1970 has been termed 'the Mexican miracle'. There have already been many documented studies on the sombre concomitants of the miracle (e.g. Navarrete, 1970; Hansen, 1971). Here I want to point to some of its consequences in the countryside.[4]

1. There are now more people in the rural areas, and for most of them the situation is one of critical poverty. Even if the relative numbers active in agriculture diminished from 70 per cent of the total active population in 1930 to 49 per cent in 1970, absolute numbers increased by 3.8 millions. Land distribution never caught up with population increase: there are at present about four million landless people in the countryside (Flores, 1970: 16). On the other hand, the expansion of large-scale urban industrial manufacturing and the aggressive penetration of urban commerce into the countryside have led to the virtual extinction of many crafts and domestic industries, so that alternatives for rural employment other than agriculture are scarce.

2. The alternative of urban work for landless country people is far from being able to absorb the demand. The inability of the industrial and service sectors to provide enough employment is manifested in the fact that only 30 per cent of the natural increase of the rural labour force actually went into those sectors. On the one hand, most country people lack the formal training which modern industry and services seem to require, as the Cárdenas plans for rural technical education were discontinued. On the other hand—and this is the essential factor—Mexican industry has been oriented towards a capital–intensive, not labour-intensive, pattern of growth (Benítez Zenteno, 1972: 566–8). Furthermore, some 30 per cent of Mexican industries are working under capacity (Suárez a, 1972). To provide employment for the new contingents of labourers who keep arriving on the market it would be necessary to create half a million jobs every year—an impossible feat under the present conditions (Trejo, 1970).

3. The 'polarity' (a term frequently used by economists which is also in the official vocabulary) of Mexican agriculture is an important consequence of these two processes. The increasing volume of agricultural production has not prevented a decline in the rate of growth of the agricultural product per active head from 2.9 per cent in 1940–49 to 1.2 per cent in 1960–69. This means that the vast majority of

agricultural workers produce very little and earn even less (Reyes Osorio and Eckstein, 1974 : 22–4). The Ministry of Hydraulic Resources.[5] has estimated that only 20 per cent of the land was producing a surplus, whereas 27 per cent was at bare subsistence and 53 per cent did not even reach subsistence level—as most of the smaller plots, which are also non-irrigated, provide less than 100 days' work for their holders in a year. Another official estimate is that only 4 per cent of agricultural workers hold more than 250 hectares per head, but they receive 46 per cent of the national agricultural income (Suárez, 1972).[6] Among this 4 per cent one should note the existence of a handful of 'super farms': in 1960 there were 2,035 top farms (in productivity and extent), the average size of which was 2,231 hectares (Tello, 1968: appendix, table 13). These large-scale farmers contrast strikingly with the vast majority of *minifundistas* (very small holders) and landless labourers (Table 5.1).

Table 5.1

Social composition of the economically active rural population in Mexico: 1950 and 1960 (thousands of persons)

Occupational category	*1950*	*%*	*1960*	*%*
Small landowners: less than 5 hectares	803.8	16.66	719.3	11.7
Landowners: more than 5 hectares	339.5	7.03	483.5	7.86
'Entrepreneurs'	14.2	0.29	19.3	0.31
Ejidatarios	1,430.9	29.66	1,523.8	24.79
Paid labourers	1,378.3	28.57	3,289	53.52
Family labourers (unpaid)	857.2	17.76	110	1.79
Total	4,823.9	99.97	6,144.9	99.97

Based on Alejo, 1970: 111–2 (the original sources are the national censuses)

Interpretations of the phenomenon of 'polarity' in agriculture vary. Government spokesmen tend to see it in terms of the classical 'modernisation' theory: it is a question of technical backwardness and administrative inefficiency in the poor rural sectors which could be eliminated through development aid projects. Others (Stavenhagen, 1969: 12 ff., Flores, 1970) point out that the agrarian reform programmes should be updated so that problems such as demographic growth, international competition brought about by technology, and lack of market organisation, could be dealt with. It is argued that to think in terms of 'mere land distribution' more than fifty years after the revolution is patently absurd (García, 1967: 21 ff.). It is also stressed (again by Stavenhagen, 1969: *passim*; also by Esteva, 1975, and, in a more forceful way, by Warman, 1975 and 1976) that the poverty of the many is not independent of the prosperity of the few, in so far as the wages earned by rural labourers are extremely low and the agricultural sector has in fact transferred capital to other sectors of the economy. It has been estimated that, from 1942 to 1961, 2,500 million *pesos* were transferred through private banks from the rural to other sectors, and that 3,600 million *pesos* were transferred through the price mechanism (Reyes Osorio and Eckstein, 1974: 30–1).[7] It is significant in this respect that a proper system for fixing official minimum prices (*precios de garantía*) for agricultural products was not established until 1953, and even then it covered only a handful of 'basic' products—maize, beans, wheat. This has prevented the price of these products from collapsing—but also from rising. The average rate of increase for non-agricultural prices has continued to be much higher than the average rate for agricultural prices. In 1970 the return from selling a ton of maize bought less than half the industrial products it could have bought in 1953. (Officially, inflation was kept to a minimum between 1953–1970. In fact agricultural prices remained stable, and for some products even declined, whereas industrial prices rose steadily.) In this situation the big farmers, with access to credit and sophisticated technology, and oriented towards specialised cash crops, are able to switch production according to the most favourable market conditions, or even to withdraw from agriculture temporarily or permanently so as to direct investment to more profitable ventures. With less room for manoeuvre, the small holder is comparatively defenceless.

Another crucial factor in the impoverishment of small agriculturalists is their inadequate and disadvantageous access to the agricultural market itself. The lion's share of agricultural profits (more than 40 per cent, according to some estimates: cf. Alejo, 1970: table III–6) goes to middlemen, in transport, retailing and, above all, storage costs. Again, the big farmers tend to be better equipped to circumvent the middlemen by creating their own transport and storage systems. Yet another factor is the general corruption among agrarian authorities, which produces a variety of effects: from failure to apply expropriatory laws to latifundia (and this relates to the fragmentation of village holdings owing to land scarcity for villagers) to inefficient handling of credit, agricultural extension and rural services in general.

The undeniable impoverishment of the rural population leads writers such as Bartra (1974) to detect a trend towards proletarianisation of the rural labour force concomitant with the vanishing of *minifundia* and the triumph of capitalist agricultural concerns as the overall organising force in the countryside.[8] In this respect, official statistics are ambiguous (Table 5.1): if the number of small private landowners seems to be decreasing, the number of *ejidatarios* has in fact increased. On the other hand, the sharp distinction made in the census statistics between 'paid labourers' and 'family labourers' is false, as anyone who has lived in rural Mexico can testify. In addition, the statistics ignore women, adolescents, children and old people who, classified as 'non-active', are seldom in fact disengaged from family economic activities.

In contrast to Bartra's point of view, others (e.g. Warman, 1973, 1976) argue that the phenomenon of 'polarity' reinforces the distinctiveness of peasant organisation which, although interacting with the capitalist organisation of *neolatifiundia*, generates and perpetuates its own patterns of social action and develops along its own lines.

Rather than continuing this argument at an abstract level, it is necessary to specify the conditions under which concepts such as the 'peasantry' and 'capitalism' may throw light on the problems of the Mexican countryside. Within this perspective I intend now to examine a number of empirical phenomena which elucidate the manner in which the general trends described above manifest themselves in the Morelos Highlands. Local-level research may in turn help further to clarify and reassess some aspects of the overall trends.

New communications, new technology, new necessities

Sugar, loans and roads. In 1938 President Cárdenas founded a sugar refinery in Zacatepec—a town in southern Morelos—and encouraged the revival and growth of the sugar industry in the whole state, so that advantage could be taken of new world market opportunities. The Zacatepec refinery was organised as a government-sponsored (and government-controlled) co-operative, but private refineries, such as Oacalco, and others which would emerge afterwards, were also to participate in the plan. The activities of these refineries (even though not all of them have been successful) have since been protected by the government, as sugar was again to become one of the main national export products.[9] In 1944 a federal law decreed that holders of suitable land in the vicinity of *ingenios* or refineries should co-operate to increase sugar-cane production. Irrigation came back to many lowland fields. Thus the Morelos Highland *ejidatarios* with irrigated plots in the valley began to cultivate sugar cane to be sold for cash exclusively to the nearby refineries of Oacalco and Casasano.

Two other major changes occurred in the same period. Firstly, people were given access to small loans from government banks to develop commercial agriculture. The most important of the credit programmes was one designed to promote the growing of rice in the irrigated *ejidos*. Secondly, the government opened a broad dirt road from Yautepec and Oacalco eastward, which makes a junction with the Mexico City–Amecameca–Cuautla highway, passing through the communities of Pantitlán, Tlayacapan, San Andrés, Totolapan and Atlatlahucan[10] The success of the road was such that the slow and unreliable railway service lost most of its custom.

The new commercial crops of rice and sugar cane put money in the pockets of the *ejidatarios*. The road opened the way to lorry-owning middle men who took other regional products to wider markets and brought all sorts of commodities into the villages. Paraffin stoves, leather shoes, tailored clothes, cutlery and transistor radios became standard household equipment, symbols of prestige and economic success for those who could afford them. People's diet extended to tinned foods, various types of wheat bread, pastas, and bottled soft and alcoholic drinks. Agricultural production became increasingly oriented to the accumulation of cash.

Technology and tomatoes. The transformation of local agriculture took place in this context of increasing monetisation. An important technical innovation was the introduction of steel ploughs, again around 1940. These were lighter than the old wooden ones; they could be drawn by mules instead of oxen, and ploughing thus progressed much faster.[11] Steel ploughs could also penetrate deeper and could be used in fields where the soil was previously too hard to cultivate. In addition, farmers began to use chemical fertilisers. Fertilisers were firstly introduced by the sugar refineries and the rice programme. Seeing their effectiveness, people began to apply them to other crops as well.[12] But perhaps the most important technological innovation of all occurred in the late 1950s, when modern methods of cultivating tomatoes spread throughout the region.

The growing of two varieties of tomatoes (red tomatoes or *jitomates*, and green tomatoes or *tomates de cáscara*) had been traditional for centuries. But it was a chancy business, for rain and insects frequently took a heavy toll. In 1955 an Italian immigrant living in Cuautla rented some land near Atlatlahucan and started cultivating red tomatoes, with impressive results. His methods involved the use of selected imported seeds, fertilisers, insecticides, and an arrangement of sticks and wires like that used in vineyards—to keep the plant growing above ground so that the fruit did not rot in the humid soil. Some villagers started to imitate him, and soon people noticed that production increased tenfold by comparison with the traditional techniques. In Tlayacapan a local teacher who was also a part-time farmer applied himself to the task of instructing his fellow villagers in the Italian's methods, having tried them out for himself. The new technique was also diffused through informal social networks.

A whirlwind of change. The introduction of this important new commercial crop was helped by a marked rise in the price of tomatoes during the 1950s and early '60s, and by further improvements in communications with the outside world. During the 1950s, more unpaved roads were opened, giving access to practically every village. In 1960 the main road was surfaced, as in 1971 were the roads to Tlanepantla and Nepopohualco. Yet most small settlements could still only be reached by dirt roads which during the most severe rainy periods in July and August are almost impassable to vehicles. In 1965 the construction of the Mexico City–Cuautla motorway, which passes immediately below the Morelos Highlands, made travel easier and faster. Regular bus services were established, and, thanks to the motorway, it became possible to get to Mexico City by bus from any of the four municipal seats in about two hours. A bus journey to Cuernavaca, the state capital, takes about an hour; to Yautepec and Cuautla, some thirty minutes.

Once changes had begun, they accelerated rapidly. Electricity was introduced in 1960, and succesful tomato growers and merchants—and there were now local merchants—were among the first to acquire television sets and record players from Cuautla retailers. The government educational system expanded slowly but steadily (See De la Peña,1980, Appendix IV). In the late 1960s a Mexico City paper mill started to exploit the northern pine forests. Since then the mill has hired local labourers for short periods, thus providing another source of income. Communications were improved in 1970–71 by the installation of a public telephone booth in each municipal seat. In 1974 a highway cutting across the Ajusco was inaugurated. This made it possible to reach Xochimilco from Tlalnepantla in thirty minutes—the same journey that would previously have taken an expert muleteer fourteen hours. Newspapers and magazines arrive regularly in the towns and, even if few people actually read them, the main national and international news is common property. Television, radio, magazines and newspapers contribute to the spread of propaganda in favour of mass-produced goods and the adoption of a 'consumer society' style of living. A wave of weekend tourism—and some tourists are buying plots and houses in the area—completes this panorama of change and cash orientation.

It is important not to see this phenomenal invasion of outside influences as the cause of social change, as if the local population were simply shaped by it. In this connection, it should be observed that the expansion of sugar-cane production did not go as well as had been expected, nor could it provide dry-season employment for everyone. If the area devoted to sugar cane over the country as a whole practically doubled from 1935 (87,000 hectares) to 1945 (164,000 hectares), the annual rate of expansion in the following ten years was only 4 per cent (J. de la Peña *et al.*, 1955:68). This period saw a new market recession due to the end of the war and a return to normality in international trade. The Mexican currency was devalued in 1950, and again in 1953, bringing the economy to a halt—including the sugar refineries. Thus the Morelos lowlands have never again been able to absorb as many seasonal labourers as they did in earlier centuries. For the Highlanders this meant active involvement not only in cash-crop production on their own plots but also in migration, educational careers and non-agricultural activities. This involvement will be examined in the following sections.

It should be added, finally, that all these innovations have not brought social harmony. Conflict and discontent continue. During the 1950s the federal agrarian authorities would not visit the Highlands without a military guard. The opening of the Cuautla motorway in 1966, which passed through irrigated *ejido* land (private estates seem to have been carefully skirted), stirred anger—and the army turned up again, just in case anything should happen. Many people had their plots partially or even totally obliterated by the motorway. They received compensation of 1,000 *pesos* per hectare—20 per cent of which was discounted to cover administrative costs, documentation and tax stamps. The price of irrigated land at the time was about ten times as much. But the protests were not very loud. Fresh in people's minds was the memory of a lowland leader, Rubén Jaramillo, who had been killed together with his family by the federal police in 1959 because he was held responsible for active—and violent—protest against abuses from the Zacatepec refinery.

Population trends and migration: Morelos and the Highlands

Table 5.2 provides data for a comparison between twentieth-century population trends in the Morelos Highlands and in the state of Morelos taken as a whole. At the turn of the century the growth of the resident population which had occurred in the nineteenth century was slowing down, because land monopoly on the part of the *haciendas* favoured out-migration. During the revolutionary period (1910–21) the population decreased by 50 per cent in the Highlands and 42 per cent in the state, owing to people being killed or leaving, to low fertility rates brought about by hunger and separation of spouses, and to high infant mortality.[13] Then, from 1921 to 1970, the population of the Highlands rose by 300 per cent, and that of the state by almost 500 per cent.

In both cases a factor which should be taken into account to explain the post-revolutionary increase is the relative improvement in medical facilities. Mortality rates, especially of infants, have steadily dropped. Chronic epidemics—including influenza, tuberculosis, smallpox, typhus and malaria—have been eradicated by intensive, government-sponsored vaccination campaigns (Holt-Buttner, 1962: 38–40). Of course, medical facilities are by no means available equally in urban and rural areas. In the Highlands, nearly 50 per cent of the deaths recorded in the municipal files are still of children under three years of age. (A common cause of infant mortality is dysentery, which is in turn caused by drinking *jagüey* water.) However, medical graduates doing their social service period have visited the municipal seats regularly since the early 1950s and, in emergencies, there are the public hospitals in Yautepec and Cuautla.

On the other hand, demographic variations cannot be regarded as independent of the material and social conditions of production and reproduction (Godelier, 1975; Meillassoux, 1977). Local population growth has depended upon the persistence of small-scale, labour-intensive agriculture. This productive activity is organised and carried out through the manipulation of differentiated relationships, mainly kinship relationships. In this sense, demographic patterns are closely interwoven with patterns of kinship and domestic reproduction. It is significant that—according to the 1970 census—women over forty years old living in the four Highland *municipios* have had six children

Table 5.2

Population trends in the Morelos Highlands *municipios*

	*Atlatlahucan**		*Tlalnepantla*		*Tlayacapan**		*Totolapan**		*Total*		*Total for the state of Morelos*	
Year	*Number of people*	*% increase or decrease*	*Number of people*	*% increase or decrease*	*Number of people*	*% increase or decrease*	*Number of people*	*% increase or decrease*	*Number of people*	*% increase or decrease*	*Number of people*	*% increase or decrease*
1900	–	–	2,242	–	4,618	–	2,266	–	9,126	–	160,115	–
1910	–	–	2,412	+7	5,608	+21	2,847	+25	10,867	+19	179,594	+12.6
1921	–	–	809	–66	3,157	–43	1,362	–52	5,328	–50	103,440	–42.4
1930	–	–	989	+22	4,000	+26	1,896	+40	6,885	+30	132,068	+28
1940	2,139	–	1,418	+43	2,421	–	1,703	–10	7,681	+11	182,711	+39
1950	2,462	+15	1,721	+21	3,037	+25	2,345	+37	9,565	+24	272,842	+50
1960	3,193	+29	1,948	+13	3,728	+22	2,642	+12	11,511	+20	385,264	+42
1970	5,167	+61	2,627	+35	5,235	+40	4,059	+55	17,088	+46	616,119	+60

*Atlatlahucan became an independent *municipio* in 1933. Its population before 1940 is included in the records for Tlayacapan and Totolapan

Sources: Holt-Buttner (1962: 31); National Censuses of 1960 and 1970

per head (six successful pregnancies) on average. This figure is extremely high, considering that childless women, as well as widows and divorcees, are included in the total. In fact, many couples make a conscious effort to beget as many children as possible. The ideology is that the strength and health of a male are reflected in numerous offspring, and that God helps those families who accept His will. But people are also acutely aware that children represent hands for the field and security for old age. It is only recently that young couples have begun to question these ideas and to use contraceptive techniques.

We shall come back to kinship and family at the end of the chapter after looking at some related themes. Now we turn to another factor which is crucial in analysing demographic trends: the changing rate of in- and out-migration, which in turn highlights social and economic changes in so far as the latter condition migration flows. At the state level, it is possible roughly to relate population growth to four forces of attraction for migrants. The first was land reform in the 1920s and '30s. With all its limitations, it provided a reason for many villagers to go back to their abandoned fields. The second was tourism and its concomitant job opportunities: in the 1940s Cuernavaca expanded as a fashionable place for foreigners to retire to and for urban dwellers to spend their holidays. In the 1950s and '60s tourist development schemes have emerged everywhere where land and water are available, or can be snatched from villagers. Even traditional peasant towns such as Tepoztlán—and, to a lesser extent, Tlayacapan—have been invaded by weekenders, hippies, American investors—and anthropologists on vacation.

Two other forces of attraction for migrants since the 1950s have been industrial development and labour-intensive commercial agriculture. Industry began to emerge in lowland urban areas, especially in Jiutepec and Emiliano Zapata, two *municipios* adjacent to Cuernavaca, as a result of national policies of industrial decentralisation.[14] Thus sugar was no longer the only industrial product in Morelos. As for commercial agriculture—onions, alfalfa, tomatoes, rice, sorghum—it too broke the sugar monopoly and even invaded large tracts of the Highlands. Season after season, poor labourers from Guerrero and Oaxaca come to work in the harvesting of these products, and many of them stay.

On the other hand, migration patterns in Morelos cannot be described simply in terms of a field of forces of attraction, whereby labour pours into those areas and occupational sectors where it is in demand. Immigration occurs in a complex interplay with emigration and circular migratory processes. These processes take different forms at different periods of time, and people have taken part in them in various ways. In the Highlands, the pace of in-migration has been slower than in the state as a whole (De la Peña, 1980: Table 5.3), and out-migration has adopted a variety of forms. An examination of the Tlayacapan sample may lead to a better understanding of these particular processes.

Migration patterns in the Highlands

I shall argue that to treat migration as a uni-directional, or single-cause, phenomenon would lead to obvious distortions. There are few studies of migration in Mexico, and some of them (e.g. Muñoz, Oliveira and Stern, 1972; Kemper, 1976) take into account only the rural–urban trend—thus ignoring circulatory phenomena—and tend to reduce migration to a pattern of push–pull causation. However, migration data collected with open-ended techniques are especially apt to show that rural and urban settings, peasant and industrial worlds, are not sociologically discrete phenomena. As Garbett and Kapferer have argued, 'migratory behaviour should be understood in terms of the location of the migrants within a field of relationships which may be extended across both rural and urban contexts', and they go on to say that 'the variation observed in migratory behaviour is directly related to a migrant's different involvement in various sets of relationships located in this field' (1970: 184). In the same vein, Roberts suggests that 'the content and scope' of a field, which 'often cuts across the physical boundaries between provinces and cities, ... is more important in influencing the orientations [of migrants] ... than spatial location' (1974: 208).

The examination of data which now follows similarly attempts to throw light on the importance of rural–urban linkages (not only for migration but for Highland social life in general) and on the heterogeneity of the social field in which Highlanders participate.

Categories of Migrants. The Tlayacapan sample (eighty household heads, males) can be classified as follows:

(*a*) People who have never migrated (thirty-four);

(*b*) 'Revolutionary' migrants (two): people who joined the army during the revolution but came back afterwards and stayed.[15]

(*c*) Returned labour migrants (thirty-two);

(*d*) Returned educational migrants (four);
(*e*) In-migrants (eight).

I shall comment on each of these categories and also on the conditions related to their emergence.[16] I shall also refer to another category which for obvious reasons is not included in the sample (although many households provide data about it): out–migrants who at present live outside the Highlands.

People who have never migrated do not seem to be concentrated in any particular age group; but twenty-eight out of thirty-three had access to some type of land, and thirty-two out of thirty-four were in occupations that were either purely agricultural, or else mixed (a combination of agricultural and non-agricultural activities) (see Table 5.3.) In contrast, a comparatively higher proportion of returned labour migrants do not participate in agricultural activities or have access to land. Moreover, most returned migrants with mixed occupations got their land *after* they had spent some time outside the area. Similarly, the fathers of non-migrants are more heavily concentrated in agricultural activities than the fathers of returned labour migrants. In fact the importance of family labour in Highland agriculture—which will be analysed later—tends to some degree to preclude the migration of members of agricultural families for long periods. On the other hand, a family tradition of non-agricultural activity (such as commerce, tailoring, carpentry) helps a man to find an urban job.

Types of returned labour migrants. However, labour migration is a term which covers a rather heterogeneous phenomenon. Various types of migration activities were registered in the sample. First, there is semi-skilled urban work. If functional literacy is not a significant factor differentiating migrants from non-migrants (Table 5.3), it certainly conditions the chances of finding an urban job: with one exception, all cases of semi-skilled urban employment corresponded to functional literacy. (The exception is a Tlayacapan trained carpenter, who spent more than ten years in a Mexico City workshop before 1930.) There is a single case of skilled urban employment, and the man—an electrician—is also literate. Unskilled urban jobs—carriers, sweepers, construction workers—could be found without any formal schooling, though nowadays many enterprises demand a school certificate even for the lowest posts. For both semi-skilled and unskilled urban labour migrants, the favourite destination was Mexico City.

Table 5. 3
Returned labour migrants and non-migrants: some comparative traits

	Migrants (*N* = *32*)	*Non-migrants* (*N* = *34*)
A. *Present occupation*		
Agricultural	11 (34%)	21 (67%)
Non-agricultural	9 (28%)	2 (6%)
Mixed	12 (38%)	11 (32%)
B. *Functional literacy*	19 (60%)	18 (53%)
C. *Father's ocupation*		
Agricultural	17 (53%)	29 (88%)
Non-agricultural	12 (38%)	2 (3%)
Mixed	3 (9%)	3 (9%)
D. *Wife's place of birth*		
Morelos Highlands	22 (69%)	29*(90%)
Other places	10 (31%)	3 (10%)

* Two were single. Source: Tlayacapan sample

Long-term agricultural labour migration within Mexico—excluding seasonal labour in the lowlands—needs no formal training or qualification, but is rather uncommon: there are only four cases in the sample, and two of them migrated before land distribution. The four cases remained within Morelos. On the other hand, agricultural work in the United States (*bracerismo*) became popular in the 1950s. At that time it was possible for *braceros* to get a permit from the U.S. government to cross the border. [17] A Mexican agent for American farming companies visited Morelos around 1950 and took many men back with him to do harvest work in California, New Mexico and Arizona. Usually these men went to and fro every year. In the 1960s both the Mexican and the American governments limited the flow of *bracero* labour. On the other hand, the supply of Mexican workers had increased enormously and it was becoming very difficult to find work. Most Highlanders stopped going to the States, and the few who continued to do so went illegally. For most people the *bracero* experience was not pleasant: they maintain that they were overworked, crammed into barracks, and generally treated badly.

Educational migrants. It should be noted that, out of the nineteen literate migrants, only two went beyond primary school, and even those two did not finish *secundaria* (secondary school). Lack of formal schooling is hardly surprising in a region where educational facilities are very scarce (see De la Peña, 1980, Appendix IV). However, there are people whom I call educational migrants who got out with the explicit purpose of furthering their education. Four heads of household in the sample are returned migrants of this type. Three are teachers: rural teachers often manage to obtain jobs in their places of origin. The fourth went to Cuernavaca to study *secundaria* and *preparatoria* (pre-university courses). Before finishing the latter he got a job in a car plant. About the same time he married a Tlayacapan girl. It is less than an hour from Tlayacapan to Cuernavaca by bus, and shift work often leaves him free for two or three days together, so he decided to go back and live in his home town, where he can help his father and brothers in their agricultural tasks, and where his wife's family live.

Migrants living outside Tlayacapan. Most educational migrants do not go back to their home town, especially those who go to university. Their specialised training—they argue—does not have a proper utility in a rural area like the Highlands; on the other hand, their earnings in the city are in fact much higher. Sixteen households in the sample (20 per cent) reported having members of the family living away; these are thirty in total, and can be divided into various categories:

1. Six are young men of twenty to twenty-five years who are about to obtain university degrees. (It should be remembered that, technically speaking, a Mexican degree requires the completion of a dissertation in addition to written final examinations.) All of them have studied for their degree in Mexico City. Their subjects are civil engineering (one), law (one), accountancy (one) and electronics (three). All, except the law student, are working: the three electronics engineers are with the national telephone company in different parts of the country, and the other two work for private enterprises in the capital.
2. Six are teachers (three males and three females) in different parts of the country.
3. One girl is studying to become a teacher at the teachers' training college in Tlaxcala.
4. Five people (four girls and one young man) studied for semi-professional careers (book-keeping, nursing), and have corresponding jobs in Mexico City (two), in Puebla (one) and Cuernavaca (two).
5. Five others (two males and three females) studied one or two years of *secundaria* and then got semi-skilled jobs (clerks, shop assistants, mechanics).

6. Six women are migrants by marriage: their husbands left Tlayacapan.
7. One girl is a domestic servant in Mexico City.

Several points should be noted about these thirty people. All except two (one married women and one clerk) are under twenty-five. Seventeen (over 50 per cent) live in Mexico City, and the rest (except a girl who married a Tlalnepantla farmer) in large cities, usually in state capitals. Migration is directly related to education in the majority of cases. Even several married women migrated because their husbands are teachers or have semi-professional jobs.[18] Finally, it is noteworthy that very few women go to the cities as domestic servants. This offers a contrast with some other rural areas in Mexico. One could hypothesise that the reasons are the importance of females within the local family structure, and the economic opportunities available in the Highlands.

From the 1970 age structure of the Highland *municipios*, one could hypothesise that—apart from the educated migrants—the vast majority of those who leave in search of wider horizons eventually come back. The only notable irregularities are in the age cohorts fifty to fifty-four and fifty-five to fifty-nine—which correspond to the revolutionary years—and in the age cohorts twenty-five to twenty-nine and thirty to thirty-four—which correspond to the Cárdenas recession. (De la Peña, 1980, Appendix V.)

In-migrants. Kinship networks seem to play an important role in the process of in-migration. Of the eight in-migrants in the Tlayacapan sample, two had consanguineal kin in Tlayacapan, one had affinal kin (before coming), and one had both affinal and consanguineal kin. A fifth had affinal kin in Tlalnepantla. But the perception of money opportunities also played a role. All of them came to Tlayacapan after 1950, and they all said that economic betterment was a reason for coming.

Four have done well. Alejandro Bernal, with six years of schooling, the son of an Ozumba shopkeeper, inaugurated the bus station at Tlayacapan, opened a restaurant for the drivers, and manages the local post office and the public telephone booth with the help of his wife. He is also the local agent of an American tomato-buying company. Jesús Garrido, with six years of schooling, the son of a Mexico City mason, trained as a smith and a solderer with the national oil company at Minatitlán (Veracruz), and opened a smithy and mechanical workshop where he works at his forge and repairs doors, windows and ploughs as well as tractors, cars and lorries. Apolinar García, with no schooling, the son of a Cuautla mason, drives his own lorry and carries produce from Highland settlements to Cuautla and Mexico City. Modesto López, with nine years' schooling, the son of a Puebla farmer, trained as a mason in Puebla, is busy repairing roofs and walls, and is also building some new houses.

Of the other four, one is an elderly man who manages a small pharmacy and also receives an allowance from his son, a doctor in Mexico City who did his social service in Tlayacapan as a medical student. The other three manage to make ends meet as agricultural labourers.

The Oaxacos. Even if none of them appears in the sample, it is necessary to mention another important category of in-migrants: the so-called *oaxacos*. These are people from the Mixteco sierra (an abrupt range which cuts across the southern states of Guerrero and Oaxaca) who come every year for two or three months at the end of the rainy season to work in the harvesting of tomatoes—which requires an intensive labour input—and also in the cutting of sugar cane in the lowlands. They dress in rags. Their language is Mixteco—though most of them also speak broken Spanish. They come from several different villages, always in groups of siblings or cousins. Some women come with them; they do the cooking and work in the fields—in contrast to Highland women, who rarely participate in agricultural tasks.

Oaxacos move from one village to the other as jobs become available. They sleep under the arches of town halls or market places, or in churchyards. Some of them manage to get a room with the poorest families. But most Highlanders do not want to have the *oaxacos* near their houses. Local people openly despise them: 'the Indians' (*los indios*) they call them. The local stereotype of these 'Indians' is that they are drunken, dirty, lazy, dangerous, sexually promiscuous ('they bring one woman for ten men'), and ignorant. In fact the presence of these seasonal workers is disadvantageous for local agricultural labourers: *oaxacos* increase the level of competition for local jobs, and help to keep wages low. On the other hand, without their seasonal help local labour would probably be insufficient to meet the demands of the tomato harvest.

The dynamics of migration

One of the rare Central American migration studies which departs from the uni-directional, urban-centred model is Nutini and Murphy's essay on the Tlaxcala–Pueblan area (1970). They discovered three types of circulatory migration (weekly, daily, seasonal), which implied on the part of the migrants the establishing, conservation and manipulation of relationships dispersed over a wide social spectrum—from the rural household to the urban industries. Their essay provides important arguments for rejecting certain prejudices about urbanisation and industrialisation, namely that these processes imply 'disorganisation' of the rural milieu and destruction of the extended family. On the contrary, Nutini and Murphy demonstrate—as does Van Velsen (1960) for Central Africa—that urban and industrial involvement on the part of villagers implies both a careful organisation and a strengthening of extended-family links, because the latter are a crucial factor in determining the success of circulatory labour migration. They convincingly argue that male absence from the village would be impossible without the set of concrete services that wives and children receive from a wide kinship universe. However, it is precisely their focus on the extended family which leads these authors to refrain from analysing the full range of social relations in the migratory field, and to view the causes of circulatory migration as merely, on the one hand, the need to supplement agricultural income (inducing them to leave) and, on the other, the fact that migrants 'have a social and economic view oriented towards the community' (Nutini and Murphy, 1970: 87) (inducing them to return). Thus they fail to consider migration as a process in which people adapt to a changing context through the strategic use of different resources in a field.

In order to understand the processual nature of migration in the Morelos Highlands, one should look, firstly, at the long-term changes. After the revolution, economic stagnation in the Highlands, from the point of view of job opportunities, was no worse globally than in other rural areas or in the cities. However the cities offered scope for certain types of semi-skilled activities which the rural areas did not. Later, when the communications network improved, and the volume of cash circulating in the Highlands increased, activities of this nature became possible in the region. During the 1950s there was the option of migrating to the U.S.A., though its value as an alternative declined later when paid agricultural labour on cash crops became available locally.

The development of cash crops made it worth investing in agriculture. Thus earnings from migration could be channelled into land and technology, and also into buying labour. This in turn has enabled many people to return from urban employment whether as entrepreneurs or as labourers. In addition, a context for the in-migration both of entrepreneurs or traders (such as Alejandro Bernal and Jesús Garrido), and of labourers, permanent or seasonal (such as the *oaxacos*), has been created. A further factor is education–the existence of opportunities for more formal schooling than that offered by traditional local schools. These opportunities arise not only from the general expansion of the educational system and the creation of government scholarships for the training of rural teachers, but also from the villagers' new ability to meet the expenses of educating their children—and their willingness to do so.

Secondly, one needs to look at the differences in access to resources, including the degree of involvement in crucial relationships. Higher education is an unevenly distributed resource: usually it is limited to the children of the local elite. Similarly, the acquisition of certain manual skills (carpentry, for instance) tends to be related to particularistic family connections. Belonging to a family with a history of migration may provide a man with vital external links which can be strategically activated for migration purposes. The return of a migrant is also conditioned by the existence of certain types of local networks.

All these elements may be illustrated by the biography of a returned migrant.

Rolando Campos

Born in Tlayacapan, 1930. His father farmed a small plot of his own, which did not produce even enough to feed the family. Rolando was the third of eight siblings. He started working at the age of seven, taking care of the neighbour's four cows. He took them to the hills for pasture in the morning, and collected them in the early evening. At the same time he attended the local school where he completed three years. At ten, he worked in the fields with his father. But his father went to live with another woman, and Rolando's mother and brothers were in danger of starvation. His mother's cousin lived in Mexico City, and invited Rolando to work with him as an errand boy. The cousin was an electrician, so Rolando learned the trade. In 1950 he married a Mexico City girl and got a job with the national oil company, which sent him to Yucatan. There he perfected his trade, but never got a permanent post in spite of his ten years' service—because permanent posts, he reckons, were for the friends of the union bosses. Rolando never lost contact with Tlayacapan, as he visited his mother at holidays. In 1960 a friend informed him that there was a man willing to 'sell' an *ejido* plot; would Rolando be interested in buying it? Rolando was interested, so he got a 0.9 hectare irrigated field, after paying the necessary bribe. He decided to stay in

Tlayacapan, where he farms the land, and also operates as a freelance electrician. He is often hired for specific tasks by the Oacalco refinery, or by Cuautla and Mexico City enterprises. There were periods (shortly after his return to Tlayacapan) in which he sought work farther afield. For instance, in 1962 and 1963 he spent several months in Texcoco and Puebla. He has managed to keep active a wide social network outside Tlayacapan, through his wife's family and through friends from the oil company. It is through this network that he gets information about occasional jobs. But he also has an important local network in Tlayacapan through participation in ritual associations. In 1962 he sponsored the Niño Dios *fiesta*, and in 1969 the Lord of Chalma *fiesta*. He has taken part in Tlayacapan religious pilgrimages to Chalma since 1952. Even when he lived in Yucatan he managed to come most years. In 1967 he was given a plot of communal land (half a hectare), and in 1968 he became a member of the Communal Land Committee. He possesses two houses—which he has bought—one for the family, and one for renting out. Recently he has been trying to found a new *mayordomía* (ritual association) for a small shrine of Our Lady of Guadalupe which he built outside his house.

Comment As Rolando's case shows, migration is much more than a mere flight from rural poverty: it implies 'a strategically organised move . . . [which] includes the acquisition of education (or training), the access to help and accommodation in the place of migration, the access to capital and other resources' (Roberts,1973: 238). In addition, the case shows the economic constraints that often apply in urban and industrial centres. If opportunities for permanent employment in industry and urban services are critically scarce, it is strategically important to keep open the option of going back to the rural areas. At the same time, this scarcity of employment in cities is related to the fact—which is fairly common in Third World urban economies (Roberts, 1973: 245)—that the most successful migrants are those who create their own employment opportunities: as independent craftsmen, carriers, commercial entrepreneurs. Such independent ventures can be substantially supported by involvement in rural activities, which provide an economic basis for bad times. Such is the case with Rolando's involvement in agriculture on his *ejido* and communal plots, which allow him to carry on his freelance trade without excessive pressures.

Poverty and the money economy

Household expenditure. Poverty is an observable fact in the Highlands. Households may keep their expenditure to a minimim (see Tables 5.4 and 5.5) and yet still at times be unable to meet basic needs. Assuming a farmer has a small plot (half a hectare) that yields enough maize for family consumption,[19] and that he has no rent to pay on his house, he still needs to earn about 600 *pesos* a month to support his wife and children without severe hardship. (This supposes that the family practically never eat meat.[20] In a way, bottled soft drinks are more important than meat, given the local water scarcity.) To bring in this

Table 5.4
Minimum cash expenditure per adult (1971) (in *pesos*)

Food	*Weekly*	*Yearly*
Milk	2.00	104.00
Bread	7.00	364.00
Maize milling	1.40	72.80
Vegetables and salt	5.00	260.00
Bottled soft drinks	4.00	208.00
Chocolate and coffee	2.00	104.00
Rice and pasta	2.00	104.00
Eggs	7.00	364.00
	30.40	1580.80
Clothing		100.00
Shoes and sandals		50.00
		1730.80

Based on interview data collected in six settlements

Table 5.5
Minimum cash household expenditure (1971) (in *pesos*)

	Weekly	*Yearly*
Food (two adults and three children)	121.60	6323.20
Soap	2.00	104.00
Powder soap	4.00	208.00
Cooking oil	5.00	260.00
Fuel oil	6.00	312.00
Electricity	3.00	156.00
	141.60	7363.20
Clothing (two adults – three children)		400.00
Shoes and sandals (two adults – three children)		200.00
		7963.20

Based on interview data collected in six settlements

much money he would have to work practically every day as a hired labourer—local wages being about twenty *pesos* a day in 1971—which is out of the question, unless he goes elsewhere. Thus he has to accept lower living standards. Alternatively, he may use part of his little plot to grow a cash crop. At the same time, many families (in 1970, about 30 per cent), have no access to land at all, so that even maize has to be paid for.[21]

Nor are medical services free. Proper medical attention is available only in the towns of Totolapan and Atlatlahucan. In each there is a health centre supported by federal funds with a full-time nurse attached to it. But the centres are concerned mainly with maternity care, including attendance at childbirth, the provision of food to pregnant women and breast-feeding mothers, and the distribution of baby food. For other things people have recourse to the medical graduates who do their social service in the area.[22] Yet the graduates are not present all the year round, are usually confined to the four municipal centres, and often do not take up residence in the area but commute from Cuautla or Cuernavaca. What happens in the end is that most people have to use the transport facilities now available and look for medical attention outside the region. But where? The free public hospitals of Yautepec and Cuautla are good but often crowded. Besides, admission depends on the recommendation of the municipal authorities. *Ejidatarios* who grow sugar cane for the refineries are entitled to use the Cuautla social security clinic. But these *ejidatarios* are only a minority in the Highlands. And they complain of poor treatment in the clinic.[23] Thus, more often than not, Highlanders resort to private medicine—which is expensive: a surgical operation may cost 10,000 *pesos* or more. A small private clinic opened in Tlayacapan in 1971.

On the other hand, local traditional medicine—herbal cures, shamanism—is not expensive, and it is often efficient. But more and more people now prefer Western medicine.

The value of the land. There is thus a universal need for money: it is essential for survival, for medicine and for other basic necessities. Most important, involvement in agricultural activities itself demands the use of cash, to buy implements and fertilisers, and to pay the cost of transport to urban markets. As a result the relationship of human beings to the land tends now to be conceived in terms of monetary utility. A high local demand for land has been created, and with it rampant inflation in its price. Holdings which are near to a road, particularly, have shot up in value. In Tlayacapan, for instance, every plot is relatively close to a tarred road and for that reason is considered much more valuable than, say, many holdings in Totolapan which can only be reached by a two- or three-hour mule journey.[24] A four-hectare plot near Tlayacapan fetched 100 *pesos* in 1930; in 1940 it was sold for 400 *pesos*; in 1971 the present owner was asking 50,000 *pesos* for it. The demand for land—as has been said—also affects the formally 'inalienable' *ejidos*. Further, there have been one or two cases in which the *ejido* authorities themselves promoted the sale of a plot, forcing the holder to be a party to the transaction.[25]

Patterns of renting land have been affected as well. The old system of sharecropping has practically disappeared. People who let their land are no longer interested in its productive capacity, only in its cash value. Similarly, those who rent it from others are interested in expanding their farming and with it their access to finance. Rent is therefore paid in cash, and usually in advance. It varies according to the value of the crop to be grown by the tenant. Thus on one hectare for growing maize it would be about 600 *pesos*, but if the same field were to be used for tomatoes it would be 1,500 *pesos* or more. The renting of *ejido* plots has also become common practice, despite its illegality.

Who are the people who let or sell their land? Old people do, if they are childless. So do those who are in urgent need of money to pay for the treatment of illness in the family. Yet, however important medical expenditure may be, the business of selling and letting would not be stopped simply by providing a free health service. The primary reason behind it is to be found in the commercialisation and capitalisation of agriculture, which results not only in land and labour becoming commodities, but in cash itself becoming a commodity—a commodity in strong demand—as well. Farmers who at a given moment of time do not have the money needed for production inputs sometimes borrow from pawnbrokers. If the harvest is jeopardised by bad weather, or the return diminished by market fluctuations, the only way of repaying the debt is frequently to sell land. The poles of social inequality, manifested in the prosperity and expansion of some farms, and the impoverishment of others, are too often, in the last analysis, two faces of the same coin.

Occupational alternatives to agriculture

Of the Highland population registered by the 1970 census as economically active 86 per cent are engaged in agricultural pursuits (see

De la Peña, 1980, Table 5.7). These include farmers (41 per cent), paid labourers (30 per cent) and unpaid labourers (15 per cent). Farmers are those who control and work a plot of land, be it private, *ejido* or 'communal'. Since many farmers control more than one type of tenure, the distinction between private owners and *ejidatarios* has become largely irrelevant. The category of 'unpaid labourers' refers to people working on the family farm without receiving proper wages. 'Paid labourers' are those who receive a fixed wage for their work. It should be noted that most of these paid labourers are not permanently attached to a particular farm. This is because the small size of most holdings and the techniques of cultivation in use only demand the hiring of labourers at certain points in the yearly cycle. Thus labourers move from farm to farm, and compete among themselves for employment.

The census figures contain at least two gross inadequacies. First, as already mentioned, the distinction between 'paid' and 'unpaid' labourers is held to be a clear-cut one, whereas in reality it is not. Most paid labourers do actually co-operate in family farming when there is a family plot. Second, women are absent—and we shall see that women play a very active part in economic life.

But the census data point to the underdeveloped character of non-agricultural activities. This is due not only to lack of natural resources other than cultivable land, but also to a series of social factors which it is convenient to examine.

Animal husbandry. I have already mentioned the poor quality of winter pastures, and the difficulty of watering herds during the dry season. It should be added that Atlatlahucan and Totolapan were the only settlements properly endowed with pastures in the agrarian reform programme. However, a number of people in every settlement have traditionally raised some cattle. For instance, in 1937, there were eighty-three out of 392 Tlayacapan family heads who raised some cows and goats.[26] Later, in 1950, an epidemic of aphtosus fever (foot-and-mouth disease) ravaged the herds, and further discouraged the development of cattle raising as a significant occupation. Nowadays, a few people keep three or four cows to supplement their diet. The little milk these cows produce is drunk by the family. Cows are also useful as they can be sold for cash in an emergency. In addition, in every settlement there are one or two producers and sellers of beef and milk who keep from ten to twenty head each. The only proper byre in the region is in the town of Atlatlahucan; even this has the ordinary type of cattle which does not produce much milk.

Many families in the mountain *pueblos* of the *municipios* of Atlatlahucan and Totolapan keep small herds of sheep and goats, which produce meat and milk for the household (see De la Peña, 1980, Table 5.8).

Pigs and chickens are found in many households, but again they are kept for consumption rather than for commercial purposes.[27] Lack of water prevents the large-scale raising of domestic animals—particularly pigs. In addition, setting up a poultry farm or a piggery demands considerable initial investment. Given the absense of credit for these activities, it is only someone with money from other sources (a rich farmer, for instance) who could initiate this type of business. And there is a lack of technical knowledge.[28]

The exploitation of timber. The pine forests which grow in the communal land of the municipality of Tlalnepantla are under the legal control of the federal government, which is interested in preventing deforestation. Thus, local people have to ask permission from federal inspectors—who visit the area regularly—to fell trees. The number of trees that may be felled is limited.

In 1944 the federal government allowed a Mexico City entrepreneur to install two lumber mills in Tlalnepantla. The town was to receive a fixed rent of 5,000 *pesos* a year because the mills were on its communal property. However the rent was not paid until the end of the ninth year—45,000 *pesos*, which were invested in the building of a school. The lumber mills closed shortly afterwards—for unknown reasons. Then the Tlalnepantla people applied for a permit to exploit the forest commercially: they planned to organise a co-operative of wood cutters and resiners. The permit was refused. They were told that the timber now formed a part of an area exclusively reserved for a paper firm in Mexico City (Loreto-Peña Pobre). In a letter to the president of the republic in 1961, the villagers gave assurances that they could and would exploit the timber sensibly and asked that the paper firm's monopoly on the forest should cease.[29] The letter was never answered.

In 1968 the paper firm installed a lumber mill near the town of Tlalnepantla, and signed an agreement with both the municipal and the federal government by which the firm exploits 5,000 cubic metres of timber every year. In exchange, the firm pays 25,000 *pesos* (5 *pesos* per cubic metre) annually to the *municipio*, and commits itself to plant a certain number of new trees. The lumber mill functions during the summer, so it is only during this period that some villagers (about fifty) are employed as woodmen and carriers.

The major immediate benefit to the *municipio* of Tlalnepantla has been the improvement of the communications system. A substantial part of the municipal timber earnings has gone to finishing and paving the road which links Tlalnepantla to the main regional roads.[30] In addition, the paper firm constructed another road through the communal property, primarily to serve the lumber mill; however, it is also used by lorries transporting agricultural products from the fields. The rest of the municipal earnings has been spent on building a new town hall. Perhaps in future years the municipality may decide to invest its money to provide new permanent sources of employment.

Thus, although the regional forests are a source of money, their exploitation has not meant an alternative to farming. The entrepreneurs who sell timber in urban markets are not drawn from the local population, mainly because the quota of trees for felling is exhausted by the paper firm.[31]

Industry, crafts and services. Two per cent of the active population are categorised by the census as 'industrial workers'. Several youths work in factories in Cuernavaca, and commute to and from their communities daily. Opportunities for employment in Cuernavaca are, however, limited, and they are further restricted by educational requirements: employers tend to demand that employees have gone beyond primary school. Several men work in the neighbouring sugar refineries as permanent employees. These refineries also employ seasonal labourers to cut and transport the cane.

The category also includes maize millers—one or two in most villages, five or six in the towns—masons, and women who run small sewing shops. The latter face increasing competition from clothes imported from urban markets. Other traditionally manufactured items—such as tiles and mudbricks—are also being displaced by imported finished goods.

The only local manufacture which survives in spite of all is pottery. There is still a persistent demand for ceremonial pottery for religious festivities, both within the region and in more distant communities (Amecameca, for instance).[32] However, this demand is not elastic enough to allow for expansion. For this reason, potters do not earn much more than agricultural labourers, and the proliferation of this activity would further reduce their income.

Lack of demand also limits the proliferation of masons and maize millers. Millers earn good money with their electric mills (an innovation introduced during the last decade), but to install these requires some capital.

Commerce, trades and services occupy 12 per cent of the active population recorded by the 1970 census. Small-scale commercial activities often overlap with farming. Included in this category are shopkeepers, schoolteachers, butchers, smiths, musicians, government employees (police constables, municipal secretaries, tax collectors), barbers, and electricians. A traditional service is the carrying of water from the communal deposits and from distant springs into the villages. This activity is also being displaced by the lorries coming from Yautepec and Cuautla to distribute bottled, purified water.[34]

As noted, women are virtually not included in the census figures. They usually declare themselves to be devoted to household activities, and are ignored by the census enumeration. Women seldom work in the fields, even during the harvest. They are nevertheless very active. In addition to rearing children—and they do have many—and cooking—and Mexican cooking is very laborious—women carry the midday meal to the men in the fields. They are also in charge of looking after domestic animals and back-yard fruit trees. Once a week they go by bus to the Pantitlán channel, where they spend the whole day washing clothes, bathing and exchanging information. Most important, women are often engaged in small-scale commerce, which is a constituent part of the farming enterprise, as well as a significant means of supplementing the household economy. This point requires some further examination.

Commerce and the transformation of local markets. The bulk of the agricultural production in the Morelos Highlands is nowadays sold for money, and a good deal of it goes to urban markets. This could have provided a basis for the emergence of a local category of hauliers and middlemen. However, lack of capital to buy lorries limits this opportunity. In addition, hauliers operating on a purely regional basis would be idle most of the time. The management of a haulage enterprise usually involves an extended network of rural and urban contacts providing information and customers, and access to a variety of products throughout the year (Long, 1972). These requirements are better met by people living in Mexico City, or in larger intermediate towns (such as Cuautla and Amecameca), or in inter-regional commercial towns (such as Tepalcingo and Ozumba). Moreover, many urban wholesalers who buy products from the region have their own hauliers. Some products, like sugar cane and rice, are bought directly by the promoting agencies (the refineries, and the Ejidal Bank).

The less distance a product has to travel to the market, the less the cost of transport, and the greater the profit for the producers. Thus farmers depend, if possible, on local demand. This is perhaps the main reason for the persistence of local open markets (*tianguis*), where women sell part of the produce of the family farm. Being free from agricultural labour, women can go to several different towns throughout the week. Each town has its own market day; this facilitates movement and exchange of products. However, bartering does not exist, and the purpose of the transactions for the seller is to obtain cash, even a small amount. For the buyer, the main advantage of open-market transactions is the possibility of getting lower prices through bargaining in a situation of intense market competition.[35] For many agricultural products, the limits to bargaining vary according to the season, and also according to the type of relationship established between vendor and buyer: a kinsman or a patron obtains better prices.

The open market also fulfils important functions for social relations and the informal spread of information, and this is why men regularly go to the market for an hour or two. It is there that people exchange news on urban prices for agricultural products, technology, and the advantages or disadvantages of new crops. Labourers try to find potential employers. Chains of gossip develop, and political issues are discussed. Small villages do not have an open market in the proper sense of the word, but some vendors visit the communities from time to time and display their goods in a conspicuous place.

The dynamics created by the circulation of money and its universal need are related to the emergence and expansion of a category of specialised vendors, who resell goods and primary products. Many professional merchants from outside the region come to the open markets to retail goods bought in cities (clothes, shoes, leather objects, ironmongery), on a small scale. They visit markets not only in the Highlands but also in the rest of Morelos and many villages in the states of Mexico and Puebla. No local villagers have become full-time roving dealers, but many (especially women) have become small retailers within a locality. Retailing is not confined to the market place; many households store soft drinks, sweets and cigarettes, as well as coffee beans and home-grown fruit, to be sold to customers in the neighbourhood. Girls and married women from the village go down to Tlayacapan and Atlatlahucan to buy bread from the baker and resell it in their own village.

In so far as local demand permits, there has also been a development of larger shops providing work not only for women, but also for the men of a household. There are between five and ten of these stores in the towns, and at least one or two in most villages. Some storekeepers are relatively well-off farmers who have decided to invest their gains in an enterprise perceived as less risky and less arduous than agriculture. Their merchandise is usually miscellaneous; they try to include everything likely to meet a demand, however small. The larger shop in Tlayacapan includes salt, sugar, ground coffee, soft drinks, paraffin, tinned food, ropes, razor blades, scissors, stationery, asprins, various spices, ribbons, blankets, saddles, cooking oil, brooms, cloth—all randomly displayed on the shelves. Other shops also sell alcoholic beverages, and operate as public houses.[36] More specialised shops, dealing with only one kind of merchandise (such as fertilisers, or medicaments), have only begun to emerge after 1970. All the shops obtain merchandise directly from factories or large wholesalers in Yautepec, Cuautla, Cuernavaca and Mexico City, which send supplies regularly. Durable consumer goods—stoves, televisions, refrigerators—are bought directly in Cuautla or Yautepec. Well-to-do housewives prefer to do regular shopping in those towns. It should be noted that the introduction of industrial manufactures is another function of the *tianguis*, as elsewhere in Mexico (Veerkamp, 1977).

Finally, an account of the commercial activities in the region must include mention of the Lent fairs. These events, though interwoven with religious symbols and idioms, are important occasions for commercial exchange (see Chapter VII). Regional participation is most conspicuous in two Highland fairs at Atlatlahucan and Tlayacapan, as well as in those held in the lowland towns of Cuautla and Yautepec, and in Amecameca and Tepalcingo in the state of Mexico, the latter being the second largest fair in the country (see Azaola, 1976: 60–3). All sorts of agricultural and craft products flow to these fairs from the villages, to be sold in great quantities. In turn, villagers buy clothes, shoes and other industrial products: an important function of the fairs is the large-scale distribution of urban manufactures in the countryside (Bonfil, 1971: 202).

Family and household: structure and process

The two salient facts about Morelos Highland kinship which are immediately obvious to any ethnographer are the extreme scarcity of unmarried adults (both males and females), and the predominance of nuclear-family households. For instance, an electoral census of the adult population in the town of Tlayacapan reveals that only 15 per cent of the males and 12 per cent of the females are single, and that 56 per cent of households are nuclear families (see De la Peña, 1980, Tables 5.9 and 5.10). There is also a significant number of households (25 per cent) formed by extended virilocal families.

The predominance of nuclear families seems to have deep historical roots. The Aztec rulers demanded that tributes be paid by the male head of a household, and it was he who had citizen's rights, including the right to land. There were further regulations which prevented the emergence of multi-generation households. 'The expansion of the joint family was limited because a married man who was a dependant did not remain so once he himself had a married son. We must assume that the family usually broke up before this occurred' (Carrasco, 1964: 208). Colonial rule reinforced this pattern: a married man had the right to a *barrio* house and the use of the land for himself and his family. This probably caused few problems during the seventeenth century, when population pressure for land did not exist. The increase in population during the eighteenth century must have caused frequent disputes—such as the one described in Chapter III and, perhaps, the growth of joint households. The nineteenth century land monopoly could have had similar effects, but this is mere speculation, for I do not have data on the use of land *within* villages in the nineteenth century.

What we do know is that between 1920 and 1960 it was no problem in most settlements to find a cheap piece of land within the village perimeter and to build a small *adobe* house, or to rent one. Of course, the ancient stone houses of 'the rich' cost more—and nowadays it is the richer merchants and bigger farmers who live in them. After 1960 housing started to become scarce, because of population growth, but even so the rent for a medium-size house (two or three rooms, kitchen and back yard) in 1970 was between 50 and 100 *pesos* per month, depending on the state of the house. The monthly rent of a recently built house—made of brick and cement, with a cement roof and latrine facilities—might be between 200 and 300 *pesos*. Weekend tourism has now started to inflate rents.

Nuclear family, inheritance and development cycles.

The availability of housing is thus a factor which permits a young man to set up a separate household when he gets married. The land distribution programme, which gave most adult males the opportunity of undertaking independent farming, is another factor conducive to nuclear family independence. The pattern of partible inheritance is still another factor. To understand this pattern, and its variations, it is convenient to examine the case of the Rojas family (Figure 1).

Narciso Rojas, Rosendo's father, distributed his privately owned land among his sons, and left his house to his daughter Antonia, who never married and always lived with her parents. In turn, Antonia left the house to her brother's son, Alfredo, who went to live with her when her parents died. Rosendo inherited one of the fields and bought two more plus a house with the help of his wife's coffee business; he also managed to get two *ejido* plots. Thus, he left one piece of cultivable land (either a private field or an *ejido* plot) to each one of his five sons, and divided the house among his six children, on condition that Tomasa, his childless, widowed daughter, was allowed to live there. In fact, Dámaso also lives in this house with his family. The Rojas expect that, when Tomasa dies, Dámaso will buy his brothers' shares in the house.

Certain features of the Rojas case are common to many families in the region:

1. Land is divided among males.
2. Women do not inherit land, except when they have no male siblings; but they may inherit a house, or a part of a house.
3. The parental house may be divided among all siblings, or else may go to the one who stays in the house keeping the parents company. This is either a spinster or a widowed daughter, or the youngest child (still nicknamed *xocoyotzin*, in Nahuatl), who is usually the last to marry.

These rules, as well as the normative model sketched below, apply to most agricultural Highland families during the period 1930–60. Total inheritance rarely occurs until the father dies. However, it is customary for a father to divide his property among his married children before his death. In fact, a father is expected to give his sons some land soon after they marry—as a sort of marriage settlement. As unmarried sons are considered to be *hijos de familia* ('family children')—i.e. youngsters who are looked after by their parents, and do not have any significant personal expenses—they do not get any inheritance until the father dies. This being so, it is important for a youth to get married as soon as possible. However, his father will discourage him, especially if he has little land. In this connection, it is interesting to note that males marry at an older age than women (see De la Peña, 1980, Table 5.9)

The family development cycle is thus one in which every nuclear family household generates similar units. Extended virilocal families

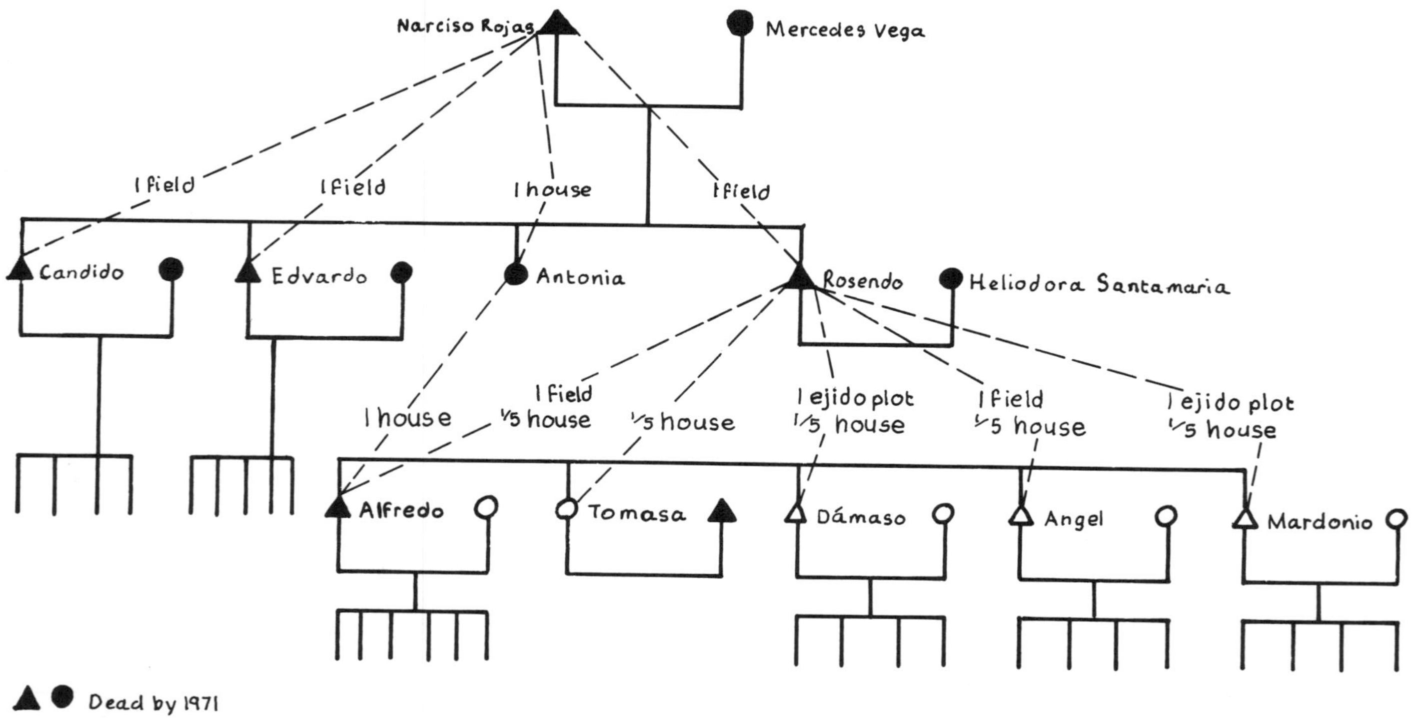

Figure 1. The Rojas family inheritance.

exist at a moment within this cycle. A young man takes his wife or girl friend to his parents' house for a variable period of time. During this period—which may never end, if the youth is the only son, or the youngest, i.e. if there are no more brothers who get married and bring their wives—he works on his father's farm and shares household expenses. In fact, he and his bride are considered as 'family children'. But the youth is expected to obtain some money of his own by occasional wage labour—the sugar-cane harvest in the nearby lowlands is important in this respect—or labour migration. There are two factors which cause the couple to leave the parental household : the birth of their own children, and the marriage of other brothers. 'You cannot live with so many people in the same house', is the explicit reason for the split. The presence of several married women within the same household may provoke disputes over household management and child-rearing, or else jealousies between sisters-in-law over the mother-in-law's preferential treatment. Management of a farm also may become difficult when several adults are involved. Thus, the father helps the son to find a house; if there is money available, the house is bought, otherwise, it is rented. Sometimes it is the girl's family who provides the house. Less frequently, if the youth's family has a house with a very big orchard or back yard, a new house is built on it.

The household split does not imply a significant weakening of family relationships. In the first place, the married son may continue working with his father, though usually his father gives him a piece of land to work independently. But this may be very small (half an hectare), so a certain dependence on the father's land persists. Second, the son knows that the land he has been given may be only a part of his future inheritance—especially when his father has a significant amount of land to divide. Eventually, the father lets everybody know what the exact inheritance will be for each child. 'Otherwise', an old man told me, 'your children would fight among themselves after you died, and your soul would not rest in peace.' The father deliberately refrains from giving everything to the sons, 'lest they become oblivious of their parents'. Third, a pattern of reciprocal help in agricultural tasks develops between father and sons, and between brothers. It is normal that certain tasks of the agricultural cycle, such as sowing or weeding, are planned in common. Fourth, family relationships are a precious resource for migrants. For instance, a migrant who goes to the United States for a season leaves his field, animals, wife and children in charge of his parents. The parents-in-law also help the migrant's family. This pattern is not a new one, because the Highlands has a long tradition of labour migration. When seasonal labour in the lowland *haciendas* meant that people spent several days at a time (or even several weeks) outside the village, there was a need for extended family support to take care of wives and children. Nowadays, transport facilities permit one-day return trips to the lowlands; but the prevalence of long-distance migration still demands a similar kind of extended family support. Even people who spend several years outside the village, and marry outside, like Rolando Campos, tend to maintain extended family networks as a strategy for possible return. In this sense, the situation in the Morelos Highlands is similar to that of Puebla–Tlaxcala described by Nutini and Murphy (1970).

The viability of household organisation. It is important that the normative model just described is not seen as a rigid structure somehow persisting over time for reasons of cultural inertia, but as a stage in a process. The type of household just described became viable because of certain environmental conditions which were created by the revolution and land distribution, and by the relationships which exist between Highlanders and external actors. Similarities with the past are due to the existence of similar conditions in the past, and not to the existence of a timeless structure. A way of testing this supposition is to examine how the structure changes as its conditions are modified.

The situation after land distribution was one in which land was—paradoxically—both scarce and available. An example of this was the fact already mentioned that, while some people were trying to use even hilly terrain for cultivation, others were abandoning their plots and moving away. Land did not produce much; its produce (maize, beans, chillies) was to a large extent for local consumption and not market oriented; production was not highly valued in the market; the value of the land as a commodity was rather low. A man could buy land relatively cheaply, or get an *ejido* plot for a small sum of money.[37] Then, he could plan to bequeathe his sons a piece of land each. Of course, the plots would not be very big: two hectares per son is, for local standards, a reasonable legacy. In turn, the sons would see their father's land as an important resource enabling them to start their farming careers, and would follow the rules of household co-operation.

However, population increase, as well as a growing orientation to cash production, has begun to change this pattern. Land prices have risen considerably, especially after the expansion of tomato cultivation took place in the early '60s. Land for sale is difficult to find and, even when found, it is expensive. As distribution stopped after 1941, the number of *ejido* plots has not multiplied. It is nowadays not infrequent to find fathers who have only a small piece of private land (of, let us say, two hectares), or else a single *ejido* plot, to divide among two or three sons. In these cases, the whole family—even after the sons have married—sticks together, at least for a while—as far as agricultural tasks on the family plot are concerned. (They may, or may not, share the same household.) The father runs the farm, and divides the earnings among his sons and himself. Divided among several nuclear families, earnings are not even enough for subsistence. Wage labour has to be undertaken and, in some cases, becomes the main activity, while the family plot is just a supplement. The plot might be sold in a crisis, or else when a good offer is made. This often happens when the father dies, and most frequently when both the father and the mother are dead. Poor farmers do not bequeathe their single *ejido* plot, or their single piece of private land, to the sons, but to the wife, so she may keep the unity of the family and the farm. But this unity crumbles without proper leadership—and a woman is not always able to exert it. Disputes over what is to be sown, about labour obligations, about individuals' financial contributions, are not always successfully settled. In the end, one of the sons buys or rents the land from his brothers, or else the land is sold or rented to someone else.

There are of course other farms which have grown and developed through successful cultivation of cash crops, but they tend to operate on the basis of wage labour, not family labour. In such cases the children often leave agriculture and go on to higher education, or start an independent business (a shop, a maize grinding mill). Or else with their fathers' help they start farming on their own account, either buying or renting land.

Both in big or small family farms, the sons now have opportunities other than fulfilling the role of 'family children'. Successful labour migration—which allows a man to save money—may lead to early independence from the parental household either by establishing a separate farm, or in non-agricultural activities. Educational migrants—obviously—are freed from the 'family child' obligations—and, frequently, also excluded from inheritance. Conflict occurs when an educational migrant comes back and claims his part of the family land. Several teachers who found work in local schools have quarrelled with brothers because the latter refuse to share the land with them. Parents tend to think that teachers, or people with an additional source of income, should not receive any land, especially when the family farm is rather small.

Strictly speaking, a single household has never been a production unit in the Highlands. Groups of households (the father's and the sons') worked their farms on a co-operative basis, although each household organised its consumption separately. (Joint consumption was very rare, for each household also had access to food and commodities through wage labour and migration; this led to differences in life-style among siblings.) With the coming of cash crops, mainly tomatoes, wage labour has become widespread. It is impossible to grow tomatoes without employing extra labour, no matter how large the extended family might be. And everybody grows tomatoes—except those who have a very small plot (0.5 hectares or less) and no capital. The latter spend more time working as paid labourers than working on their own farms. Thus, the organisation of rural labour is by and large determined by the requirements of specialised cash-crop agriculture, rather than by household needs. Of course, families are not mere consumption units: their members are productive and maintain co-operative links among themselves. But the crucial variables for the organisation of their activities are not confined to the household. As stated when discussing migration, one has to view people's activities as occurring in wide social fields, in which household norms and needs are combined with various *other* resources and relationships.

Kinship categories: consanguineal, affinal, fictitious

Kinship networks are not well-defined local organisations in which precise rights and/or obligations are allocated to people according to their position in a genealogical structure, but rather categorical relationships in which kinship is used as a social idiom to express generalised relationships of trust and co-operation. These relationships are maintained not only with parents and siblings, but also with a wide variety of bilateral consanguineal, and affinal, kin. Further, the relationships with cousins, uncles and in-laws may sometimes be stronger than the relationships with the members of the nuclear family.

The existence of bilateral kinship ties is manifested by the use of both the father's and the mother's surname (Angel *Rojas Santamaría*). Yet this does not provide a precise criterion for distinguishing between kin and non-kin. Often, a surname is shared by a large number of people who do not recognise kinship ties between themselves. (For instance, in Tlayacapan the surname *Santamaría* is shared by at least one fourth of the population. Patronymic names—González, Pérez—are also quite common.) And there are many instances of people who do not share either the first or the second surnames,[38] and yet consider themselves to be related by kinship; often, precise genealogical relations cannot be traced. In such cases, a kinship tie is acknowledged by calling a person 'cousin' (*primo* or *prima*), 'uncle' (*tío*), 'aunt' (*tia*), 'nephew' (*sobrino*), closer family relations (*hermano*: brother, *hermana*: sister, *papá*, *mamá*, *abuelito*: grandfather, *abuelita*: grandmother) are also emphasised.

Affinal kinship terms are used in the same emphatic fashion: *cuñado* (brother-in-law), *cuñada* (sister-in-law), *suegro* (father-in-law), *suegra* (mother-in-law). In addition, I found cases in which people used the terms 'cousin', 'uncle' and 'aunt' to refer to some of their affines' consanguineal kin. This happens when a special co-operative bond develops among affines, usually because of the help provided by the wife's family in the setting up of a newly-wed couple's household. When it is the bride's father who provides the house, it will usually be sited in his own locality (see, for instance, the case of Pablo Figueroa in Chapter VI). Then, the girl will be assisted in child-rearing and domestic chores by her mother, sisters and family friends. Her husband will take part in local recreational and ritual activities yet he will continue to work co-operatively with his own father and brothers. Through his mediation a pattern of frequent interaction may develop between the two sets of kinsmen. Those individuals who develop close relationships of trust may start calling each other 'cousin' if they are of the same age group, or respectively 'uncle' and 'nephew' if they are not.

All this seems to indicate that an individual's definition of his kinship relatives is very much a matter of selection—and not merely a biological given. I was able to trace genealogical connections in the third generation among individuals who denied having any kinship tie; whereas some 'uncles' or 'cousins' could only make reference to vague distant ties. (A brother cannot be denied; but brothers who have quarrelled do not see each other or speak of each other.) A person may accumulate a very large number of kinsmen in this categorical sense, too many for all such relationships to be based on co-operative bonds. Other bases have to be sought for selective kinship; one is the establishment of fictitious kinship links through the concepts of *compadrazgo* and *padrinazgo*.

Compadrazgo is the relationship established between a man (or a woman) and the godfather (or godmother) of his (her) son (daughter). *Padrinazgo* is the relationship established between a godfather (*padrino*) and a godson (*ahijado*). In the Morelos Highlands, I was struck by the frequency with which kinsmen, sometimes close kinsmen, are *compadres*, or *padrinos* to each other. In the Tlayacapan sample, fifty out of eighty household heads had *compadrazgo* relationships with persons whom they also considered kinsmen. In addition, there were sixteen cases in which the household head had more than one *compadrazgo* link with the same kinsman. Multiplicity of *compadrazgo* ties is possible because of the existence of various types of *compadres*: for baptism, confirmation and first communion. Recently it has become fashionable to have *compadres* for school graduation and fifteenth birthday parties of well-to-do girls. A man also chooses *padrinos* for his wedding (usually in consultation with his parents) and for such occasions as house-warming parties, or buying a tractor or a field.[39]

Of course, *compadrazgo* ties also develop between persons who are not kin, and are usually developed from or into relationships of trust. I found that *compadrazgo*, as well as kinship, ties existed among people of equal economic status; but they also cut across strata—they develop between farmers and labourers—and across spatial location—migrants develop *compadrazgo* ties with city people, farmers with market intermediaries, etc.[40]

'Limits to growth' and limits to trust

Population growth in the Highlands was the result of a complex set of factors—family strategies, immigration, health improvements—and caused excessive pressure upon land. On the other hand, sugar-cane cultivation in the hot country could not absorb all seasonal labourers. At the same time, the increasing dependence on urban markets created a widespread need for money. In order to cope with all these situations, people have undertaken long-distance migration and diversified local agriculture. Both strategies were partially supported by government programmes and facilitated by improvements in the infrastructure of regional communication.

Migration was a means of gaining access to some money and skills which could be invested in experimenting with new crops and other activities in the Highlands. But economic diversification has been drastically limited by scarcity of, and monopolising over, crucial local resources. As far as agricultural modernisation is concerned, it has resulted in both more wealth and more poverty. Thus, even if circulatory migration persists, permanent out-migration has also become an inevitable product of post-revolutionary 'development'. The sons of the poor leave in search of better work opportunities; more often the sons of the rich look for educational qualifications in urban areas and stay there for good. Because of their perception of limits to regional growth, prosperous farmers and shopkeepers do not expand their enterprises beyond a certain point; instead they use their money to pay for education for their children or to buy furniture and appliances.

Some of these points are illustrated in the following biography:

Isidoro Luque

Born in Tlayacapan, 1899. His father was the local tax collector in Porfirio Díaz's regime, so he and his family had to flee from *zapatista* violence. In 1919, young Isidoro—who had learned to read and write in Tlayacapan—got a job in a Mexico City bank as a clerk. He stayed there until he was twenty-five. In 1925, he ventured into the commercial world. With his savings he bought a handful of garments and went to sell them in villages near Mexico City. He was successful: there were not many salesmen on the road in those years, when violence had not yet vanished. His main operating area was the state of Mexico—where he met his wife; but eventually he came to Morelos, and to Tlayacapan. He formed the idea of opening a shop in his home town, as there was none. He did so in 1935, retailing whatever he thought he could sell. Success was slow but steady : he was meeting a demand, meagre but real—and increasing. He soon discovered that lending money at a high interest rate was also good business. He became *the* local moneylender. He was respected : together with the teachers, the priest, and one or two government employees, the Luques were identified as those 'who are somebody'. In 1950, Lorenza, the eldest of the Luque children, married young Lauro Alarcón, the son of a prestigious local teacher (Doña Leovigilda). Lauro was a successful local politician: he was at the time municipal secretary, and shortly afterwards municipal president. (At the moment, Lauro works in the District Treasury.) The Luques and the Alarcóns formed a tightly-knit *compadrazgo-padrinazgo* network. Lauro opened a shop himself with Isidoro's help, while the latter has relied on Lauro for all his dealings with the government: tax paying, permits, yearly inspections. It was with Lauro's help that he got a government permit to open a maize mill as soon as electricity came to Tlayacapan. Then, he got an exclusive concession to retail paraffin in the locality. When other maize mills started operating, he would sell paraffin only to those people who ground their maize with him. His shop flourished. He also invested in some land and cattle—which is taken care of by his eldest son, who also helps his father in the shop. His second son received a brand-new maize mill from his father when he got married. His four younger children went to study to Mexico City and stayed there.

Isidoro's career as an innovative entrepreneur includes diversification of activities, and reinvestment. However, his direct involvement in agricultural production has been minimal; his commercial ventures have not created jobs. His career also shows the importance of differentiated social relationships to gain access to resources.

The analysis of farming careers also shows that differentiated social links are necessary to cope with an economic context of scarcity, risk and uncertainty. Before proceeding, it is pertinent to examine the nature of such links, which we have also called 'relationships of trust'.

The folk concept 'trust' (*confianza*) has a variety of meanings. The category 'trustworthy people' (*gente de confianza*) refers to those one is familiar with and from whom one has no secrets. The term also implies reciprocal expectations of help; but such expectations are limited by 'common sense'. The term *abuso de confianza* (misuse of trust) refers to excessive demands from a friend. The term *confianzudo* (presumptuous) applies to people who are unduly demanding. These people inspire *desconfianza* (mistrust); therefore, their neighbours prefer not to interact with them, as it is difficult to establish a mutually satisfactory exchange of services. People who repeatedly refuse help to others are also objects of mistrust. However one should not bestow too many unexpected favours on another person (this is called *ser ventajoso con alguien*: to be 'advantageous to' someone): it suggests that too many future services are to be asked in return. In all cases, one has to rely on a more or less implicit consensus on which demands or favours are excessive and which are to be considered 'reasonable'. But this consensus should be seen as a result of the conditions in which exchange occurs,[41] and not as a pre-condition of it.

At an analytical level, the term 'trust' has to be related to the theory of social exchange. Economic exchange differs from social exchange in so far as the latter cannot be precisely quantified. It is also impossible to specify the content of social exchange in advance. Nor it is formally obligatory; thus, mutual expectations of reciprocal favours do not emerge from formal regulations but from trust among the participants (Blau, 1964: 94). When a process of social exchange has developed and persisted in a satisfactory manner for all participants, trust becomes a stable aspect of their relationship. Power emerges from imbalances in social exchange, but trust also develops from asymmetrical links as long as the chain of exchanges persists and is positively valued.

It is analytically essential to isolate those conditions in which social exchange is evaluated as satisfactory. Such conditions, and not *a priori* social norms, determine the type of social relationships to be found among individuals and groups.[42] In this connection, the examination of regional agricultural dynamics will provide important clues. Later, we shall explore some aspects of religious ceremonies which are related to those symbolic frames in which Highlanders evaluate their exchanges.

Notes

1. In a letter to Víctor Paz Estenssoro, quoted by García Cantú, 1971.

2. Most of the land distributed before Cárdenas was not the best for agricultural use. The amount of land distributed from 1920 to 1934 was nearly ten million hectares. Cárdenas distributed over seventeen million hectares in six years (see Silva Herzog, 1964: Ch. XI).

3. The Agrarian Code was modified to increase the size of lawful private land holdings and to ratify and enhance the right to *amparo* or injunction on the part of landowners. In his posthumously published testament, Cárdenas (1971: 10–11) condemned all these measures as being serious obstacles to the implementation of the 'goals of Mexican Revolution'.

4. Unless otherwise stated, the source of these data is Centro de Investigaciones Agrarias, 1973.

5. Quoted by Mexico City newspaper *Excelsior*, 10 November 1971.

6. See also De la Peña (1980), Appendix II, Table III.

7. These estimates are extremely moderate, according to the authors themselves.

8. Bartra is faithful to Kautsky's unilineal models. However, he also uses the concept of 'permanent primitive accumulation' to explain the functionality of smallholder agriculture for capitalism.

9. Cyclical crises—brought about by social conflicts and inefficient planning—have affected sugar production in Mexico in the last thirty years. There were eight refineries in Morelos at the beginning of the fifties; fifteen years later, there were only three which stand to this day: Zacatepec, Oacalco and Casasano. From 1975, Mexico has been unable to export any sugar, partly because of the demands from an expanding national market (see Del Villar, 1976).

10. A new settlement—called Kilómetro 88 because it is at that point of the highway—sprouted up by the junction of the highway and the new road.

11. The diffusion of steel ploughs in Morelos really gained momentum in the late 1950s. In 1950 two-thirds of the ploughs recorded by the Agricultural Census in the state were still of the old type (see Lewis, 1951: 32 ff.; and Ingham, 1968: 60 ff.).

12 Chemical substances used in the Morelos Highlands nowadays include artificial guano for maize; ammonium sulphate, ammonium nitrate and urea for tomatoes; ammonium and potassium sulphate, and sodium nitrate, for sugar cane; calcium phosphate, ammonium sulphate and urea for rice. Among the more widely used insecticides and fungicides are the substances called Aldrin, Manzate and Tarsate.

13. See Coy's comments on the demographic patterns of Tepoztlán (1971: 47). Holt-Buttner (1962: 29) provides a lower figure for the state population in 1920: 99,140 people.

14. See Arias and Bazán (1977) for a history of Cuernavaca industrial development.

15. Their biographies are included at the end of Chapter IV.

16. I am not classifying as *migration* the involvement of most people in work in the immediate lowlands. As the Tlayacapan *ejidos* are there, *ejidatarios* go to the lowlands

everyday. Given the present travel facilities, this journey only takes fifteen or twenty minutes.

17. Legalisation of *braceros* occurred in 1942. Mexican labour was needed 'to help with the harvesting of crops and to work in railroads because of the labour shortage brought about by World War II' (Whetten, 1948: 27).

18. Marriages between people of different Highland settlements are relatively common. A quick survey in the village of San José de los Laureles (110 households) revealed that there were three women from Tlalnepantla, two from Tlayacapan, one from San Agustín, all married to San José men. In the same situation were one girl from Xochimilco and one from Yautepec. In turn, there were San José girls in Pantitlan (one), Tlalnepantla (three), Tlayacapan (three), San Agustín (three); and also in San Andrés Ahuayucan (three), a village in the state of Mexico which maintains a ritual-exchange relationship with San José. Males usually do not undergo this type of marriage migration.

19 Normally, family consumption amounts to six *cargas* (bags) per year. These can be reaped from a plot of half a hectare using just a tiny quantity of fertiliser.

20 See Appendix II, Table 2.

21 People are conscious of social and economic differences in a rather detailed manner. The Tlayacapan water committee had to fix differential rates for 814 adult males. They discussed them one by one and the final classification was: upper category 6.38 per cent; middle category 37.78 per cent; lower category 56.75 per cent; very low category 3.07 per cent.

22. Dysentery from drinking *jagüey* water is frequent. When asked why he did not boil the water he drank, an *ejidatario* said that he would still be drinking water from irrigation channels if he got thirsty when working in the fields, so what was the use of it? Pneumonia is also frequent among children and old people.

23. *Ejidatarios* feel that they are despised because of their poverty and rustic manners, and that other social categories, such as clerks and urban workers, are given preferential treatment. In addition, they maintain that they do not receive all the medicines needed for a cure, so they have to spend money anyway. All this reinforces a preference for, and trust in, private medicine.

24. The *ejidatarios* of Tlayacapan usually go to their plots by bus.

25. See A.S.R.A. *Nepopoalco*, 1971.

26. See A.S.R.A. *Tlayacapan*, 17 November 1937.

27. To keep a pig is a way of saving money. A baby pig can be bought for 100 *pesos*; if it is reasonably fed with left-overs, after a year it may be worth ten times more.

28. A few farmers were able to benefit from a limited government loan for piggeries in 1973. But this type of loan did not occur again.

29. A.S.R.A. *Tlalnepantla*, 28 November 1961.

30. Between 1970 and 1971, total expenditure on this road amounted to about 700,000 *pesos*. A third of this sum came from the municipal funds. The rest was provided by the federal and state governments, by the firm itself, and by voluntary co-operation of the villagers.

31. The pine forests existing in the *municipio* of Totolapan have remained largely untouched, with the exception of an area near the village of Nepopualco, which has recently been exploited by Cuautla entrepreneurs.

32. Manufacture of rockets and candles for religious festivities also survives on a

very small scale.

33. Pottery is an all-family occupation, and it is practised by fifteen Tlayacapan families, all living in the *barrio* of Santiago. Clay mud extracted from hills near Atlatlahucan is transported to Tlayacapan by donkey or mule. After being pulverised and sifted, the material is thoroughly wetted, kneaded, sprinkled with sand, and put into moulds. It is fired in hand-made, wood-burning furnaces of mud and stone. This primitive method results in the rather crumbly pottery (as furnace temperatures are not very high) which is unsuitable for wider markets. Nevertheless, potters have recently been able to sell to wholesalers of Toluca City, and perhaps tourist interest in primitive crafts may lead to increasing demand. Special clay figures—some multi-coloured, human-shaped small candle sticks—are made once a year for the Day of the Death by two families of the Santa Ana *barrio* (see Rojas, 1973).

34. Water is also becoming a commodity instead of a resource of communal appropriation. People who own a private cistern sell water to their neighbours; this phenomenon is related to the decaying of most *jagüeyes*.

35. For a comprehensive description of this type of market in Middle America, see Nash (1967).

36. In a small village such as San José de los Laureles, the following items can be bought from the two local shops: evaporated milk, powdered milk, instant coffee, stationery, tinned food (sardines, chillies), sweets, biscuits, sugar, soft drinks, cooking oil, candles, matches, thread, chocolate.

37. Legally, there is only one way of getting an *ejido* plot which has already been granted to somebody else, and that is when the incumbent abandons it. But there is consensus among *ejidatarios*—and they defend this point of view in public assemblies—that a man with an urgent need for cash should be entitled to sell his plot, on the condition that the buyer lives in the same *pueblo*.

38. In the Spanish-speaking world, it is very rare to find people who bear more than two surnames, though sometimes hyphenated or double surnames are used to emphasise a particular branch of kin.

39. I found Mintz and Wolf's article on the subject extremely illuminating (1950).

40. In 1966, John Ingham applied questionnaires on *compadrazgo* ties to a set of Tlayacapan respondents. He concluded (Ingham, 1970b) that *compadrazgo* is asymmetrical, and used by poorer persons to establish patron–client ties with the rich, because godparents tend to be of a higher economic status than the parents of their godchildren, and because they do not reciprocate (i.e. they allegedly do not invite the parents of the godson to become godparents of their own children). These data should be qualified by the fact (which Ingham does not mention) that on many occasions it is the godfather himself who seeks the link. On the other hand, Ingham considers that *compadrazgo* or *padrinazgo* ties other than those established through the sacraments of the Church have secondary importance, and consequently such ties were ignored by his analysis (1970: 283). In fact some informants would say that non-sacramental *compadres* are really unimportant. But in the process of selectivity described above, all *compadres* or *padrinos* have importance, because the content of the relationship is not so much derived from explicit norms as from the exchanges that occur between two persons over a period. Ingham does not mention the aspect of selectivity. Foster's analysis of *compadrazgo* in Tzintzuntzan has a similar limit: his statistics include baptism godparents only (see Foster, 1969).

41. Larissa Lomnitz (1975) has written an excellent description of relationships of trust in a Mexico City shanty town. However, I must disagree with some points in her analysis. She puts too much emphasis on the normative aspects of exchange (p. 210), instead of focusing attention on the material conditions which both make possible and put limits to exchange. On the other hand, Lomnitz refers to *confianza* as an 'ethnographic trait' (p. 212). Such reification of the concept prevents her from isolating the situations in which the concept is to be applied. Relationships of trust imply scarcity of resources, but not all relationships which emerge in a context of scarcity are to be defined as relationships of trust.

42. In the last analysis, norms should be seen as dependent on conditions of exchange. But norms present an idealized version of the general functioning of a social group; they are frequently inconvenient for many particular situations and have therefore to be ignored or manipulated.

VI
THE DYNAMICS OF AGRICULTURE

Empirical observation must in each separate instance bring out empirically, and without any mystification and speculation, the connection of the social and political structure with production. The social structure and the State are continually evolving out of the life process of definite individuals, but of individuals not as they may appear in their own or other people's imagination, but as they really are, i.e., as they are effective, produce materially, and are active under definite material limits, presuppositions, and conditions independent of their will. [Karl Marx and Friedrich Engels, 1969: 287]

As any human process, agricultural production develops in a context of choice. A point of departure in the analysis of this process is to look at variations over time in the statistical distribution of agricultural choices, in terms of the aggregate amount of land dedicated to different crops. It is assumed that this distribution is a result of the aggregate number of decisions made by farmers. Therefore, it is possible to discover the logic of decision-making by examining the changing pattern of local opportunities and constraints for the growing of different crops. In relation to this pattern, I shall examine how different types of farmers develop strategies and meet obstacles.

The limitations on local economic expansion are also present—probably in a sharper form—in agriculture. The region presents what has been called 'the back-wash effect' (Myrdal, 1957) : environmental disadvantages (scarcity of land and water) are reinforced by cumulative disadvantages in the social organisation of production and marketing. Thus, when one speaks of constraints and obstacles, one is largely referring to a social context. On the other hand, this social context does not merely set negative limits, but also provides positive opportunities for certain types of action. This point deserves to be expanded a little.

The 'anatomy of backwardness' has often been analysed in terms of what we may call the organisational variable: 'underdeveloped' people do not organise their activities in a manner which is rationally directed to increase production and expand resources. The concept of formal rationality implies that a given set of means are functionally and systematically allocated to a certain end: in this case, to maximising economic returns. Two different models have been used to explain the emergence of rational organisation. The first is psychologically rooted: development is taken to be associated with a special type of personality. Human beings who possess this personality are motivated in their actions by *achievement*, and their behaviour consistently presents a host of corresponding traits: punctuality, efficiency, frugality, willingness to take calculated risks (see Pieris, 1969 : chs. II and III).[1] Unfortunately for them, 'peasants' are not supposed to have this sort of personality, submerged as they are in a cultural world of fatalistic values (Rogers, 1969 : 19 ff. and *passim*). The second model focuses on the social variables which hinder or favour economic expansion. The definition of these variables is a function of different theoretical orientations. Thus, an orientation germane to André Gunder Frank's view of peasantry as a dependent human group would emphasise the presence of power relations organised from the top of the social pyramid. An orientation germane to Chayanov's or Nash's concepts would emphasise internal, local variables: for instance, that household structure limits labour organisation, or that ritual diverts surpluses from investment in productive activities.

To be sure, the psychological model could be useful if associated with research which would single out the causes of different personality types—instead of taking these types as static and given—and which would analyse the feedbacks that motivations receive from perceptions of past experiences, present alternatives, and expectations about the future. Similarly, the sociological model should analyse all possible constraining links of an actor, as well as the strategic role played by his own utilisation of social relationships. Put in another way, any sound analytical model should include contextual and situational variations in factors conditioning behaviour. If actors may pursue different types of goals, then there are different ways of being rational. The aim of research, therefore, should be 'to discover the conditions under which an individual or a group may display one kind of rationality rather than another' (Kapferer, 1972: 6).[2]

Bearing this in mind, I attempt to analyse variations in agricultural behaviour at different periods and in different situations. I stress that agricultural choice is not merely conditioned by aspects and circumstances of production, but also by the farmers' participation in non-agricultural activities and by their involvement in both differentiated and non-differentiated social relationships.

Changes in the distribution of crops, 1950–1970

Census figures show that the total cultivated area in the Morelos Highlands expanded drastically from 1950 to 1960 (Table 6.1). But in fact in 1950 the summer was exceptionally dry and many fields were left

Table 6.1
Main crops in the Morelos Highlands, 1950, 1960 and 1965 (hectares, metric tonnes and '000 pesos) (current prices)

	1950		1960		1965	
	No.	%	No.	%	No.	%
Maize						
Area	3,032	90.3	4,567	67.4	4,330	64.4
Production	2,727	15.3	3,646	11.4	4,580	11.8
Value	935	64.0	2,779	38.6	4,580	17.8
Red tomato						
Area			367	5.4	1,244	18.5
Production			1,808	5.6	15,108	38.7
Value			1,690	23.6	14,960	58.5
Sugar cane*						
Area	213	6.2	357	5.2	105	1.5
Production	15,000	83.8	24,582	76.9	14,300	36.7
Value	453	31.0	1,209	16.8	750	2.9
Rice						
Area	41	1.3	29	0.4	536	7.9
Production	120	0.7	178	0.5	2,588	6.6
Value	48	3.3	158	2.2	2,980	11.7
Other†						
Area	72	2.2	1,474	21.6	506	7.2
Production	41	0.2	1,760	5.5	2,396	6.1
Value	26	1.7	1,356	18.8	2,275	8.9

Notes
* Includes first plants and ratoons
† Beans, chillies, peanuts, squash; 1960 figures include green tomatoes also; 1965 figures include green tomatoes and sorghum also

Source: 1950, *III Censo Agrícola, Ganadero y Ejidal*; 1960, *IV Censo Agrícola, Ganadero y Ejidal*; 1965, unpublished data, Dirección de Economía Agrícola, Secretaría de Agricultura y Ganadería.

idle. Had the census been taken one year earlier, the recorded area would probably have been larger. However, it is very likely that the cultivated area showed some actual expansion in the years 1950–60. First, this was the decade in which the children born in the late 1920s and early '30s (it should be remembered that there was a baby boom after the low childbirth rates of the revolutionary period) married and became independent farmers. Second, the introduction of steel ploughs provided many farmers with the possibility of expanding their cultivation. Third, the availability of new economic opportunities brought by agricultural schemes and by improvements in communications attracted agricultural investment—for instance, from returned migrants. (In addition, with these changes people needed more money and had to produce more.) Fourth, the expansion of the cultivated area was intimately related to the introduction of new crops rather than simply to an expansion of existing ones.[3]

The reasons for the increase in the total recorded value are manifold. National inflation is one. However, it cannot explain the 1,625 per cent increase in the Morelos Highlands' production value (from 1,462,000 *pesos* in 1950 to 25,545,000 *pesos* in 1965): during the same period, national price indexes rose by 100 per cent (Alejo, 1970: Table VIII-1). Maize prices per ton increased from 342 *pesos* in 1950 to 1,000 *pesos* in 1965 (an increase of 192 per cent) due to government mechanisms which subsidised the prices of basic agricultural products to ensure continuing cultivation of staple foodstuffs (Haag and Soto, 1969: 52–3, 392 ff.). The price of a ton of rice rose from 400 *pesos* in 1950 to 1,150 *pesos* in 1965. But sugar cane prices per ton only increased from 30.2 *pesos* in 1950 to 52.2 *pesos* in 1965 (not even 100 per cent). On the other hand, even if productivity per hectare had an increasing trend in all crops—mainly because of the use of fertilizers and steel ploughs—this cannot explain the rate of value increase for the total production either[4] (see Table 6.2).

Table 6.2
Productivity per hectare (1950–65) (in tons)

	1950	*1960*	*1965*
Maize	0.89	0.79	1.05
Sugar cane	70.42	68.85	136.19
Rice	2.92	6.13	4.82
Red tomato	—	4.92	12.14

Source: Table 6.1

What has really caused the rise in production value is a marked change in local farming choices which has led to diversification of crops and particularly to the adoption of tomato cultivation. In 1960, red tomatoes covered 5.4 per cent of the fields, but contributed 23.6 per cent of the total value of production. In 1965, the corresponding percentages were 18.5 and 58.5. Green tomatoes are not explicitly included in census data; but their presence may be important for explaining the relative expansion of the category 'Others' from 1950 to 1960.

The expansion of sugar cane and rice cultivation has only occurred in the *municipio* of Tlayacapan—and even there, the sugar cane has lately decreased. The reason for the lack of expansion throughout the region is simple: both crops need irrigation, and the other three *municipios* have limited access to water.[5]

However, the region seemed to present a remarkable trend of production growth, whereby new, valuable, sophisticated, export-orientated cash crops have been substituted for traditional crops. It seemed to provide an example of the optimistic official forecast for Mexican agriculture, which predicted the beginnings of a new age of rural affluence.[6] However, after 1965, the percentage of land devoted to valuable crops does not seem to have increased very much. For instance, the farmers included in the Tlayacapan sample still farm maize on about 50 per cent of their land (Table 6.3).[7] Further, in conversation they appear sceptical about the real benefits which crops such as tomatoes have for the majority of them. This raises two questions: (*a*) given that

Table 6.3

Distribution of crops in fifty-two Tlayacapan farms: 1970 and 1971 (autumn cycles)

	1970		1971	
	Hectares	%	*Hectares*	%
Maize	81.7	49.21	83.3	49.61
Red tomato	32.3	19.45	23.8	14.17
Green tomato	17.3	10.42	18.5	11.01
Sugar cane*	15.1	9.09	17.9	10.66
Rice	6.2	3.73	6.9	4.10
Others†	13.4	8.07	17.5	10.42
Total	166.0	99.97	167.9	99.97

* First plants and ratoons

† Beans, peanuts, chillies, sorghum

Source: Tlayacapan sample

Tlayacapan has an enviable position within the region in terms of quality of land and access to transport facilities, why should farmers be reluctant to depart from traditional crops? (*b*) If they doubt the real advantage of tomato cultivation, why do they persist in growing tomatoes?

I shall try to answer these questions by examining the organisation of agriculture in the general context of the Highlands. I refer first to the main regional crops: maize, sugar cane, rice, tomatoes. The conditions of cultivation for other products will be examined later. I also attempt to examine how agricultural organisation is subject to variations according to a farmer's access to resources, as well as to his differential involvement in past and present social relationships.

General patterns of cultivation

Maize and tomatoes grow both in the Highland plateaus (in any of the types of alluvial soil) and in the lowland *ejidos*. Their growing cycle lasts six months; therefore, they can be produced twice a year by those *ejidatarios* with access to irrigated plots. The main agricultural tools employed in their production are steel ploughs drawn by mules, hoes, machetes, and harrows. A few wealthy farmers possess tractors; but most people do not consider them to be a worthwhile investment for cultivating small plots of land, especially as tractors are only used in the first stages of agricultural work (i.e. to clear the land and make the furrows).[8] The more sophisticated techniques are associated with the growing of red tomatoes, which requires imported seeds, chemical fertilisers, fungicides and insecticides, bought from local retailers or, more frequently, from shops in Cuautla and Yautepec. Nowadays, however, chemical products are used on all crops.

In many aspects the cultivation of maize and tomatoes is basically similar. Work for the rainy season begins in April or early May, when the fields are cleared of the remains of the previous harvest (*solar*), and the land is ploughed and harrowed (*barbechar*). Furrows are opened just after the first rains (*surcar*) in late May, and sowing takes place immediately afterwards. Once plants emerge, the fundamental tasks of cultivation are: (*a*) spreading of fertilisers and insecticides; (*b*) keeping the furrows in good shape (*encajonar* and *despachar*); (*c*) opening channels between the furrows, thus allowing the rain water to flow freely and preventing total flooding; (*d*) weeding. But in spite of these similarities, the crops differ in one important respect. Tomatoes require a great deal of care and intensive labour to prevent disease and contusion. Maize is stronger and does not need continuous attention, though care has to be taken to prevent invasions of moles, for example.

Harvest-time for maize and tomatoes depends on the altitude of the fields. Lowland *ejidos* are subjected to higher temperatures and produce quicker; the harvest occurs in late October and early November. Highland plots are harvested in late November and early December. The cycle of work for the irrigated lowland areas which produce maize or tomatoes during the dry season begins in November and finishes in May.

Sugar cane and rice are grown only in the lowlands as they need irrigation and higher temperatures; they produce one annual crop as their growing cycles are longer. Sugar cane is sown in December, after the field has been burnt, cleared, turned over, watered, fertilised and ploughed. Irrigation is needed during the first period of cultivation (January to April), before the rains begin in May. A substantial part of the labour, both before and during the rains, involves maintaining a proper system of drainage. Other aspects of cultivation include weeding, ploughing, remaking of furrows, and spreading of fertilisers and insecticides. The *zafra* (sugar-making process) begins at the end of the following winter: after the burning of fields, the cutting continues throughout the whole spring. As they are semi-perennial plants, canes continue to grow from the same rhizomes for a period of three years, after which the field must be sown with other crops for at least the next two years.[9]

Rice cultivation is far more laborious and complex than that of sugar cane because of the technicalities involved in the supply and control of water. Operations begin in January with the building of levelled seed beds, each bed requiring a drainage system which allows water to flow continuously. The bed is flooded three times, alternating with ploughing and fertilising before the seed is actually sown. The alternate flooding and weeding takes place while the plants are growing; the level of water has to be carefully raised as plants become bigger. In May, the rice sprouts are transplanted to another field, which has already been watered and ploughed; there, flooding again alternates with fertilising and weeding. Rice is harvested in late September and October.

Irrigation is carried out using a rather primitive ditching system. Several canals built of stone, brick and cement carry the water into the lowlands from the Yautepec River and the Oaxtepec Springs. Secondary channels (*apantles*) distribute water from the canals to the individual plots. The main canals were built with government financial help and loans from the neighbouring refineries. *Ejidatarios* do not have unrestricted access to water, but each village *ejido* has a local officer or *fontanero*, paid by the Federal Hydraulic Department, who is in charge of allocating irrigation to every plot according to its particular needs. Water from both the river and the springs is pumped into reservoirs by electric engines installed by refineries. The measurement marks in these reservoirs enable the amount of water allocated to each plot to be controlled. However, for certain plots, which are situated on low-lying ground near the river, pumping is not necessary (and, therefore, the amount of water is not measured either).

The organisation and costs of production

The irrigation needed to produce sugar cane and rice is restricted to one fourth of the agricultural land of the *municipio* of Tlayacapan, and to small tracts of the other *municipios*. The use of this land is confined to a segment of *ejido* holders, and to those farmers who manage to rent a plot of irrigated *ejido* in spite of the agrarian law. Maize and tomatoes, on the other hand, can be grown on any type of agricultural land (irrigated or non-irrigated; private *ejido* or communal; granted, bought, inherited or rented). Other variations in the conditions of production relate to the type of labour organisation required for different crops, and to the type of investment.

Before examining costs of production, I want to make it clear that I attempt to present these costs as the farmers themselves perceive them. These perceptions are objectively grounded—they are far from being a figment of the farmers' imaginations; but they differ from the estimates that would be made by an economist, or an accountant. The farmer is directly concerned with how much money he needs to pay for fertiliser and labour for example. He does not estimate the cost of his own work in monetary terms—nor do I (indeed, if the labour of the farmer and his family is included in the estimates of production costs, as a strict economic analysis would demand, most farms would be permanently in the red). [10] Nor is the depreciation of tools or the rent value of land taken into account. All estimates are made for 1971 prices.

Production and costs of maize and tomatoes. Labour for the growing of maize and tomatoes is organised by the farmer himself. The initial work is done by the adult males living in the farmer's household and, to a lesser extent, by the farmer's wife and elder daughters also. At certain phases of the cycle, a pattern of co-operation will emerge between the farmer, his father, and his male siblings. However, in the case of tomatoes, which need more intensive cultivation, hired labour is also required. The differences in household composition may thus affect production costs.

Hired labour is easily available in the region. It is provided both by the local contingent of landless people (or those people with very little land) and by the seasonal migrants (the *oaxacos*) who come to the region every year. But it has to be regularly paid for in cash, and the farmer himself has to find the money for payment. The farmer also has to find the money to buy chemicals and to provide implements.[11] In addition, he is expected to give his labourers a snack at noon, so that they do not have to go back to the village for lunch.

Associated with these features of organisation we can now make an estimate of the costs of production per hectare for maize and for the red and green varieties of tomato. As an example, I take a farmer who has access to the use of one hectare of seasonal land, and who is helped by one son, his wife, and one daughter. I assume that this farmer has the necessary technical knowledge and tools, and that his land is of reasonably good quality. If he decides to grow maize, he does not need to worry about cost of seed, for this is usually available in the household from the previous harvest, or can be bought very cheaply from a neighbour. The labour required in the fields—about ninety man-days per hectare of which about twenty are employed in the harvest—can be done leisurely by the farmer and his son; the shelling and packing of the maize by the two women. Cash would only be required to buy some fertiliser and insecticide, and it would only amount to 450 *pesos*. (The need for fertiliser can be minimised by leaving some stubble after the harvest and thoroughly ploughing it in.)

In contrast, the cultivation of one hectare of red tomatoes requires an initial investment of 350 *pesos* to buy imported seeds[12] and, given the ecological conditions of the region, also an investment of nearly 5,000 *pesos* in chemicals. In addition, 1,750 *pesos* are needed to buy poles and galvanised wire for building trellises.[13] The amount of labour input needed for red tomatoes is about 400 man-days per hectare. If the labour requirements were evenly distributed throughout the agricultural cycle, the farmer and his son would be able to provide all the labour needed for cultivation (if both of them worked every day during the six months of the cycle); similarly they could do the harvesting and packing with the help of the two women of the household. However, the delicate nature of the product demands that certain tasks be performed quickly, concentrating the labour requirements of the season into four peak periods—hence the need to hire labour. First, the sowing must be done immediately after the first rains and before the ground becomes too soggy. Second, during the heavy rains of July and August, insecticide and fungicide have to be spread over the fields two or three times a week; in addition, furrows must be continuously re-shaped to avoid accumulation of water. Third, fixing poles and wires and training the plants have to be carried out as soon as the plants appear. Fourth, harvesting and packing have to be done very quickly, before the fruit is too ripe. The operation of packing tomatoes in wooden boxes is especially complicated, as tomatoes must be graded according to size and carefully arranged to avoid contusions. The wages of labourers—18 *pesos* per man-day—needed for these peak periods, therefore, raises the cost of production of red tomatoes to nearly 11,000 *pesos* per hectare (see Table 6.4).[14]

Table 6.4

Cash investment per hectare in fertilised fields of maize and tomatoes during the rainy season (1971) (in *pesos*)

Input	*Maize*	*Red tomato*	*Green tomato*
Fertiliser	430	2,700	800
Insecticide and fungicide	24	2,000	500
Seed	–	350	150
Disinfectant	–	55	55
Sticks	–	1,000	–
Wire	–	750	–
Labour for sowing	–	360 (20 man-days)	360 (20 man-days)
Labour for cultivating	–	990 (55 man-days)	360 (20 man-days)
Labour for harvesting and packing	–	2,520 (140 man-days)	540 (30 man-days)
Total	454	10,725	2,765

The growing of green tomatoes occupies an intermediate position between maize and red tomatoes in terms of complexity of production and costs. The total amount of work needed per hectare is approximately 200 man-days. Poles and wires are usually not required, as the plants are smaller than those of the red variety. The plants are also less delicate, so they need less intensive care and application of pesticides. Yet cultivation still demands hiring labourers for peak periods. In the case under consideration, the farmer, if he is growing green tomatoes, would need to hire only the labour equivalent to seventy man-days out of the total 200. The total cash investment, then, including wages, seed, and chemicals, would be about 2,700 *pesos* per hectare (see Table 6.4).

All the calculations given above assume that cultivation of any of the three products occurs during the rainy season. Irrigated cultivation in the lowland *ejidos* during the dry season permits some decrease in costs, especially for tomatoes. The reason is that the humidity can be better controlled and thus the plants are less exposed to fungi and parasites. The amount of chemicals can be reduced, and the rhythm of cultivation is slower. In these circumstances, the cash investment needed for one hectare of red tomatoes is about 9,000 *pesos*, and for one hectare of green tomatoes, about 2,000 *pesos*.

Production and costs of rice and sugar cane. The main difference between the production of maize and tomatoes on the one hand, and of sugar cane and rice on the other, is that the latter are promoted, and even partly organised, by external agencies. In the case of sugar cane, the major promoters are the two big neighbouring refineries.[15] A farmer holding a plot of irrigated land only needs a machete, a hoe, and a part of his own labour to grow sugar cane. The plot is cleared and furrowed by tractors paid for by the refinery. The refinery also provides fertilisers and money to hire extra labourers for cultivation, and, when the time of the harvest comes, organises the cutting. The farmer only has to provide about eighty man-days of work during the period of cultivation.

The production of rice is in turn sponsored by a State bank (the Banco Nacional de Crédito Ejidal), which provides seeds, chemical substances, and money for wages up to the equivalent of 100 man-days. (In total, the bank provides about 3,000 *pesos* per hectare.) The labour input in rice cultivation is practically unlimited, because the hydraulic technicalities involved can be endlessly perfected (Geertz, 1966: 31–2); however, local people usually invest about 200 man-days per hectare. The labour is organised by the farmer; but this usually does not demand any personal cash investment, as work not directly financed by the bank can be easily provided by the farmer himself and his household.

The availability of credit

It is clear then that the production of tomatoes, particularly of red tomatoes, is a special case because it requires as a necessary condition access to a large amount of cash. A farmer with a very successful career behind him may actually possess the money at the beginning of an agricultural cycle. But most farmers are not that successful, and have to get the money elsewhere. The alternatives are three: State banks, individual money-lenders, and wholesalers in Mexico City. There is, however, a form of credit, which is not termed as such, but which strictly speaking implies borrowing money from others: to withhold payment of labourers until after the harvest. This mechanism is relevant to further points in the analysis. Not all labourers accept deferred payment, only those who know the farmer well and *trust* him. This trust depends on the existence of links between farmers and labourers outside the labour situation.

The banks to which local farmers can resort are the Banco Nacional de Crédito Agrícola, the Banco Agropecuario, and the Banco Nacional de Crédito Ejidal.[16] The first two have the function of helping small proprietors, through the agency of the Ministry of Agriculture which operates in Cuautla, or directly through the bank branch in Cuernavaca. It is relatively easy for a farmer to obtain a small loan from these banks against the mortgage of his private holding, provided his deeds are properly registered, and he does not have any previous mortgage.[17] Two thirds of this loan, however, is frequently given to the farmer in the form of fertiliser of inferior quality—not very good for tomatoes—and only one third in money.[18] The total sum, with rare exceptions, does not exceed 1,500 or 2,000 *pesos*. In recent years, a number of small-scale tomato farmers have failed to repay the loans on time. Thus, bank officials are extremely reluctant to finance tomato cultivation, which is considered to be—and is—a very risky crop, so the farmers usually pretend that they want the loan for more secure products.[19]

Reasons of security also prevent the Banco Nacional de Crédito Ejidal from financing tomatoes. This bank is exclusively concerned with *ejidatarios*, and directs its resources to promoting specific crops—in Morelos rice and, to a much lesser extent, maize. The authorities of each *ejido* negotiate loans and have to see that they are used for the purpose stated by the bank, so it is difficult to divert the money to growing tomatoes. In some cases the bank refuses to give any type of loan simply because it does not have enough funds available. No wonder: the whole budget of the bank at the national level is not sufficient to finance even one third of the total number of *ejido* holders in Mexico.

In these circumstances, people who want to grow tomatoes frequently have to resort to private moneylenders in Cuautla and Yautepec—or even more distant towns; local storekeepers also advance loans. They lend cash at an interest of 10 per cent per month,[20] which means that, at the end of an agricultural season, the sum to be repaid has augmented by 60 per cent and a substantial part of the profit goes to the moneylender's purse. This rate of interest is illegal, and the moneylenders do not have any lawful means of demanding payment; but people usually pay their debts, lest they acquire the reputation of not fulfilling their obligations and become unable to get further loans.

It might be illustrative to describe a local scene where money-lending relationships are seen at work:

The moneylender and the mason

One morning in September 1971 I was strolling in the central square of Tlayacapan when a young woman, carrying a baby in her arms, asked if I knew where to find Father Humberto—one of the local priests. She explained that she wanted the priest's advice and help because some people wanted to seize her husband's house as he owed them money. As the priest was away, I offered to go with Maria—that was her name—to see whether something could be done.

First, we went to see Ricardo, Maria's husband, who was working as an unskilled labourer for a local mason. He told me that in December 1970 he had borrowed money from an acquaintance of his mother in the town of Chimalhuacán (in Mexico state) in order to pay a previous debt. This consisted of money he had borrowed from a Tlayacapan storekeeper in order to grow tomatoes. Ricardo had failed to make a profit from agriculture in 1970 and thus resorted to a new moneylender through his mother, who was born in Chimalhuacán: he paid the storekeeper with this money. As in 1971 he had no money left to invest in agriculture he let his land out to another farmer (two hectares of non-irrigated *ejido*) and took the job with the mason, hoping to save enough to pay the new debt; but so far he had only managed to pay a part of the interest. Now his creditor had come with a legal document to requisition his house. Ricardo could not go and talk to the creditor because his employer would not let him leave work; besides, he and his family could not read well enough to understand the document. I promised to look at the document and come back with the information.

Ricardo's one-room house was on the outskirts of the town; he lived there with his wife and children, his mother and his sister (Ricardo's father had died several years ago). A middle-aged woman, better dressed than the local women, was sitting on one bed, and she looked very surprised when I turned up. Maria explained to her that I was a friend of Ricardo's; that he sent his regards to her and was very sorry for not coming, but his employer would not let him leave his work. I said that Ricardo had sent me to talk to her about the matter of the debt. The woman then relaxed a little and introduced herself as Mrs Valverde. 'Look, sir,' she said, 'I'm sure that you'll be able to understand my problem and to see the justice of my claims. I've known Ricardo's mother from my childhood, and I always thought she was an honest person, so I agreed to recommend her son to a friend of mine who generously lent him some money. I have a responsibility for this loan, and now the authorities have sent me with this document to exact the money; otherwise, this house will be taken as a payment.'

I was a bit confused. 'So you're not the creditor?' 'The creditor is Mr Gonzalo Vázquez', she said. 'I'm only a friend of both parties. Look at these papers, sir, and persuade these people to pay their debt.'

Mrs Valverde then showed me a typescript saying that Ricardo and his mother had borrowed 6,000 *pesos* from Mr Vázquez and had agreed to repay within one year; they had also agreed to pay an additional 300 *pesos* every month as interest. It was signed by Ricardo, his mother and Mr Vázquez, and Mrs Valverde as a witness. The latter said, 'As you can see, the date recorded is last December. From January to April they paid their interest; but they have not paid one penny from May to September. That is why the authorities are upset.'

I said that I thought that rate of interest was illegal.

Mrs Valverde looked angry. 'Perhaps you can read this other document, sir.' The document was an official letter signed and sealed by the municipal president of Chimalhuacán, ordering Ricardo to clear his debt with Mr Vazquez *and* Mrs Valverde immediately lest the law transfer the ownership of his house to the creditors. I was obviously amazed at this letter, and Mrs Valverde smiled. 'You see,' said Ricardo's mother, 'we left the deeds of the house with them as a guarantee.'

I said that I had to consult with Ricardo about the matter, and left the house promising to return in half an hour. Ricardo's mother came after me. 'Please don't make her angry,' she said. 'She threatened to bring a lawyer and the federal police and evict us from the house. Of course she is the real creditor; Mr Vazquez only lets his name be used because he is a close friend of hers and has political connections.' I asked her to come with me to see the local authorities. In the town hall, the municipal secretary patiently listened to the story. 'Well,' he commented, 'the creditors probably bribed the municipal president of Chimalhuacán to write that letter. But the authorities of Chimalhuacán don't have any power whatsoever outside their *municipio*. So don't worry. If you want, I can have that Mrs Valverde locked in jail here in Tlayacapan for a few hours to teach her a lesson.' I sympathised with this idea; but Ricardo's mother was shocked and rejected it. The secretary shrugged. 'O.K.,' he said, 'then give her an excuse and send her away.' He turned to me, 'You speak to her. She won't dare do anything if city folk like you are witnesses of their actions.'

Moneylenders versus wholesalers. Ricardo's mother did not want to quarrel or oppose the moneylender because she saw her as a possible source of more loans in the future. Borrowing money may be a frequent necessity for a local person; and a moneylender is regarded as a kind of benefactor—even if it is generally said that professional moneylenders are ruthless bastards. Moreover, farmers know that, if they have a successful future, other less successful farmers may ask to borrow money from them. These loans among farmers and neighbours usually carry interest—though this would certainly be lower than the interest charged by a professional moneylender. This is thought to be fair, because the lender incurs the risk of not being repaid quickly. In other words, everybody tends to feel that loans should always be repaid, because that benefits everyone. It is necessary to maintain a spirit of *trust*—so that one may borrow money when needed, and also demand payment from any debtors.

On the other hand, the conditions on which Mexico City wholesalers—the third source of credit—lend cash for cultivation are not much better than the moneylenders'. Wholesalers lend money at a lower rate (2 per cent per month); but then the whole crop must be sold exclusively to them, and at the prices they fix. Sometimes wholesalers fix prices before the harvest, at the time when the loan is contracted (this is called 'to sell at the time', *vender al tiempo*); but usually tomato prices are fixed after the harvest, so that they may be fixed below their current market value.

But this leads to the consideration of other important factors affecting farming decisions, namely, the farmer's perception of the possible gains derived from each crop. This is in turn related to his knowledge of the rate of productivity for different crops; to the risk involved in their cultivation; to their market value; and to the characteristics of the market.

Expected gains and the organisation of marketing

Table 6.5 shows some crucial elements which most farmers are likely to have in mind when they compare the advantages and disadvantages of the main local crops. I shall examine the expectations of gain and perceptions of risk in some detail.

The marketing of maize. The yield from a well-fertilised maize field is between 1 and 1.5 metric tons per hectare. Local people, however, measure in terms of volume, not of weight, which means that a good harvest amounts to about twenty-five bags (*cargas*) of 100 *cuartillos* each, or 2,500 *cuartillos* of shelled grain per hectare.[21] This yield can be predicted fairly accurately; if the plants are properly disinfected and cared for, the risk of disease can to a large extent be neutralised. In addition, maize prices are very stable (1.20 to 1.50 *pesos* per *cuartillo*), due to government control.

There are a number of alternatives open to a farmer who has produced maize. First, the total crop can be easily sold to storekeepers in

Table 6.5
The Morelos Highlands: costs, expectations and perceptions for main crops

	Cost per hectare (pesos)	*Expected gain per hectare (pesos)*	*Risk from weather*	*Uncertainty from market*
Maize	450	2,500	Low	Nil
Red tomato	11,000	25,000(?)	High	Very high
Green tomato	2,700	5,500(?)	High	High
Sugar cane	–	2,000	Low	Low
Rice	–	10,000	Medium	Low

Yautepec and Cuautla. In this case, transport expenses—use of a hired lorry—amount to approximately 300 *pesos*. 2,500 *cuartillos*, sold at an average price of 1.30 *pesos* each, give a gross gain of 3,250 *pesos*; discounting 750 *pesos* for initial investment and transport, the net cash gain is 2,500 pesos.[22]

On the other hand, if the farmer does not need a large sum of money immediately after the harvest, the grain can be stored for domestic consumption, or to sell to Ozumba buyers who come to the region; or it can be sold locally over a period in the open markets of neighbouring towns. Most maize millers store maize which they buy in small quantities from local producers. Storing maize is practically equivalent to storing cash. The grain is durable and there is a constant demand for it in the region (and, indeed, everywhere in Mexico, for it is the staple food in most places). At certain times of the year (before the harvest begins), the price of maize can increase to 1.70 *pesos* per *cuartillo*.

Maize also has a number of secondary uses. The grain itself provides good nourishment for the household poultry and pigs; the leaves are used for *tamales* (a sort of maize caneloni); the stalks for mule feed; and the stripped cobs for kitchen fuel.

The marketing of red and green tomatoes. A very successful harvest of red tomatoes yields 800 to 1,000 boxes per hectare, each box weighing about thirty kg. But success is difficult to predict. Even if plants are scrupulously fumigated and looked after, the harvest can be jeopardised by fungal infestation and bad weather. Ten minutes of hail for instance, may destroy a field. If the harvest is lost, there is no way of recovering the money invested.[23] Even if the harvest is not completely ruined, bad weather may result in a decrease in the size of the fruit, which diminishes the price. The danger of disease has increased in recent years, as virus and parasites have grown more resistent to chemical substances.

Fluctuations in the market are also a source of risk. At the end of the 1950s a box of red tomatoes could be sold for as much as 120 *pesos*. By the mid 1960s, prices ranged between 50 and 60 *pesos* per box. In 1967, the market collapsed, and prices were less than 15 *pesos* per box; 1968–70 were somewhat better years (30 to 50 *pesos* per box); in 1971 the market fell again (20 to 30 *pesos* per box). Fluctuations occur not only from year to year, but also from month to month, and even from week to week. In 1971, the price of a box rose from 7 *pesos* (January) to 150 *pesos* (April), and then fell again to 30 *pesos* in late May.

The observable long-term tendency—of which farmers are very much aware—is for prices to become progressively lower. The production of red tomatoes in Mexico has tripled in the last ten years, and the national demand has not expanded correspondingly. But the farmers also know that this trend may vary from year to year, as price fluctuations depend on complicated mechanisms involving not only the national, but also the American production of tomatoes. When American fields have poor yields, Mexican tomatoes are exported to the U.S.A. The prices then rise within Mexico, and the small farmer may have a chance to earn good money. Otherwise, the Mexican market is flooded by tomatoes and the price falls.

I have referred to the circumstances of weather and price fluctuations as *risks* faced by the farmer. By this I mean that the existence of these circumstances is known by the producers, and therefore taken into account in production decisions and expectations. On the other hand, the circumstances of specific marketing transactions tend to present *uncertainty* rather than risk (in terms of Knight's well-known dichotomy) for the local tomato producers.[24] This means that farmers frequently cannot predict what course of action they will take when the harvest is ready for market, nor the result of specific market alternatives, which are as follows:

1. *Middlemen*: Once tomatoes have been harvested, they have to be sold quickly, as there are no local cold stores, or canning or bottling plants. There is not sufficient demand in the local markets, or in neighbouring towns.[25] The effective demand is thus provided by the wholesalers of the big cities—Monterrey, Guadalajara, Querétaro,

Veracruz, Puebla, and, especially, Mexico City—and by middlemen who come to the localities. They buy a certain quantity of the product—according to their own contracts with wholesalers or exporters—and take charge of transport arrangements and costs. One particular middleman, who operates as an agent for an American company (Dixon, the biggest agricultural import–export firm in Mexico), usually offers the best prices.[26]

Other middlemen offer lower prices. What should be emphasised is that producers cannot rely on middlemen to sell the whole product. The American agent only buys tomatoes of superior quality and large size, and farmers cannot be sure beforehand what proportion of their harvest is going to meet these requirements. Nor can they predict how many other middlemen will come to the region, or how much of the harvest they will buy. Thus, even if the easiest selling procedure is to wait for middlemen to buy one's crop, it is uncertain they will buy the whole crop. In addition, the transaction itself also presents a high degree of uncertainty, as farmers do not know if they are offered a fair price or are being cheated. Some farmers started telephoning wholesalers in Mexico City so that they could compare the prices there with those being offered by the middlemen. The information they received was seldom reliable. The wholesalers tended to offer a high figure in the hope that farmers would go to Mexico City and sell directly to them; but, when farmers arrived with their tomatoes, they were given a much lower price with the excuse that the market had fallen at the last minute. The farmers, having already paid for the transport, were then virtually forced to accept the deal.

On occasions, middlemen realise that local supply is abundant, and arbitrarily lower their prices even from one minute to the next. The following scene is an example of this:

Scene in the Tlayacapan central square

A big lorry arrives. The registration plate shows it is from Veracruz. A man comes out and shouts 'Who sells tomatoes?' Eight men come forward from a group in front of the town hall. One of them timidly asks what price is offered. 'Twenty-Four *pesos*', is the answer. There is an approving whisper. But other men—about twenty—have now arrived and stand by the lorry. The man from Veracruz shouts, 'What are you looking for? Do you come to pry into our affairs, or do you want to sell as well?' The newly arrived men say that they want to sell. The prospective buyer gets into the truck seat and exchanges a few words with the driver, after which he says, 'I'm sorry, boys. My fellow here has informed me that the prices have collapsed. I can only give you eighteen *pesos* per box'. 'But you said twenty-four', protests one of the farmers. 'Yes, mate, but I was wrong. The price is eighteen. Take it or leave it'. The group disperses. 'These bastards', someone whispers. 'They are trying to rob us. But, what can we do?' 'I wont sell', says another one; 'I can wait until tomorrow'. However, most people come back with their boxes and take the eighteen *pesos*.

2. *Wholesalers*: To avoid this kind of situation, farmers must look for alternatives. They can, for instance, charter a lorry and try to sell directly to wholesalers. This increases their direct costs, for they have to pay from 3 to 4 *pesos* per box for transport. In addition, the transaction is still subject to uncertainty. Most Mexico City wholesalers are concentrated in the huge market of La Merced, and they operate a cartel to keep down the rates they offer to small farmers. They also have a pricing agreement with supermarket chains and American buyers (Suárez, 1972b). The La Merced merchants know that farmers have to sell their tomatoes quickly, because the fruit is perishable, and because they have to return home to finish harvesting. They also know that a small farmer cannot afford to pay for transport to more distant markets. The farmers themselves do not know whether going to more distant markets would be worthwhile, and they have learned to distrust long-distance information. However, they cannot even be certain of selling their crop in Mexico City itself. There have been periods of abundant supply when wholesalers simply refuse to buy.

3. *Pre-sale*: One way of securing sale is to agree a deal before the harvest, or even before sowing. This frequently necessitates getting a loan from a wholesaler in the form of a contract whereby a wholesaler or middleman assumes responsibility for the sale and then pays the farmer the price obtained, less expenses (transport: 3 *pesos* per box; La Merced market tax: 3.5 *pesos* per box) and commission (3 to 5 *pesos* per box). The disadvantage of such contracts is that they never specify prices, which depend on the state of the market at the time of the harvest. The farmer is simply given receipts for the quantities he provides. After the harvest he goes to Mexico City to receive his payment. Frequently the contractor claims to have sold the product at very low prices, sometimes even lower than those offered by middlemen—and expenses and commission still have to be discounted. Thus, although this alternative eliminates uncertainty in relation to effecting the sale, it increases the uncertainty of getting a fair price.

Another form of pre-sale is to sell the crop to another farmer before harvesting. The buyer then finances and organises the harvesting and marketing. This happens when a farmer cannot afford to pay harvest labourers, or when he feels he lacks the expertise to handle the operation properly because he has never harvested tomatoes before. (Such cases are extremely rare, and usually involve immigrants.) The disadvantage is that the buying farmer's offer is normally very low, because he has to cover himself against possible market-fluctuation risks.

4. '*The Market on Wheels*': The Mercado Sobre Ruedas was created by the government in 1971, in order to prevent abuses by wholesalers and middlemen. By joining this organisation, small producers receive a special licence to sell directly to consumers. The government arranges that, on a given day of the week, a public place in Mexico City, such as a square or a street, is used as a market exclusively for producers. However, farmers can only sell a small part of their crop in this way. The market only functions once a week, and reaches a limited number of consumers. Competition within the market is on the increase: a growing number of vendors use the market and sometimes supply is far greater than effective demand. Farmers now suspect that La Merced merchants have managed to get licences for the Market on Wheels, and are flooding it with produce to discourage growers.

All these market characteristics apply to both red and green tomatoes, although the green variety has fewer disadvantages. They are less affected by the weather; and, as there is no significant demand for them on the American market, their prices show less fluctuation.

More about tomatoes: expected gains and collective selling strategies. In spite of the difficulties listed above, tomato growers tend to give rather optimistic forecasts when asked about the profit they hope to make. Their optimism is related to the high commitment to tomato production; other alternatives for cash and labour investment are deliberately excluded in favour of a risky course of action. This is only justified if the substantial investments made in tomatoes—which entail considerable effort—actually pay off. It would be 'dissonant'[27] not to expect high gains once this type of farming choice has been made. Thus, it is common to expect a red tomato yield of 1,000 boxes per hectare, and an average price of about 35 *pesos* per box. After discounting 11,000 *pesos* for production costs, and the possible expense of transport and/or commission, farmers reckon that they may get a net payoff of 20,000 to 25,000 *pesos*. However, average prices may be as high as 70 or 80 *pesos* per box—or as low as 15 to 10 *pesos*. (When it is less than 15 *pesos*, farmers do not even bother to harvest.) As for green tomatoes, an optimistic forecast would be 450 boxes per hectare at 20 *pesos* each: a gross gain of 9,000 *pesos*. After discounting 2,700 *pesos* for investment and about 1,000 *pesos* for transport and other expenses, the net cash gain would be over 5,000 *pesos* per hectare. Of course, these gains diminish if interest has to be paid to moneylenders.

Commitment to tomato growing also entails adopting a number of short-term strategies to minimise market uncertainties as far as possible. These strategies are aimed at getting information on all market alternatives. A farmer should inquire about the true Mexico City prices from other farmers who have sold their products there in the immediate past—the day before, lest fluctuations drastically modify prices. He should also find out the names of particular wholesalers who are likely to make better offers. At the same time, he should be ready to contact travelling middlemen and compare their offers with what he knows about potential offers by wholesalers. All this he should do without neglecting to organise labour for the harvest.

These multiple operations would be very much simplified if they could be carried out within a co-operative framework which would co-ordinate the efforts of a number of farmers. But such a framework does not emerge from the situation of tomato selling itself, for farmers are competing against one another. When facing a middleman, for instance, farmers do not present themselves en bloc, nor do they demand a minimum price. On the contrary, each individual tries to deal with the buyer in isolation from everybody else. However, patterns of co-operation and trust emerge among people who have, or have developed, other types of relationships: kinship, co-parenthood, neighbourliness, ritual partnership. These relationships are used to develop networks for exchanging information, and advice, about prices in different markets, names of wholesalers, the arrival of middlemen, etc. Another important function of these networks is that the members will keep an eye on a fellow's tomato field while he is away selling the produce. There have also been more direct selling strategies which have been undertaken collectively by groups of friends and kinsmen: for example, chartering a lorry to go together to La Merced and even more distant markets. A group of friends from San Agustin had the idea of renting a warehouse near La Merced, to which they could all send their tomatoes. The idea actually materialised, and it turned out to be a good one. They take turns at looking after the warehouse and contacting all the La Merced wholesalers; they then sell to the best payer. Of course, this strategy has an obvious limitation: once the tomatoes start to get even a trifle ripe, they have to be sold, no matter the price. But to have a warehouse saves time, enables the farmers to wait a day or two before selling, and slightly increases the chance of getting a better price. Yet farmers must trust the man they leave in charge of the operation: otherwise they cannot be sure that he would not be bribed by wholesalers, or misappropriate some of the money for himself.

'Lack of trust' is the actual reason people give to explain why there is no co-operative association embracing all tomato growers in the region. In some *pueblos* there are local associations which have attempted to adopt a somewhat more formal, permanent organisation than the ephemeral groups of co-operating friends. The most successful of these organisations was one which did not emerge as an economic association, but as a religious one: the Adoración Nocturna (Night Vigil), an association aimed at keeping nightly vigil during the Holy Sacrament. It was founded by a priest in Tlayacapan, and soon included buying fertilisers and chartering lorries for La Merced among its activities. Those associations solely concerned with fostering impersonal economic networks have usually been unsuccessful.

In recent years a well-intentioned, eager Cuernavaca agricultural delegate has visited the Highlands to initiate local associations. These started well because the delegate helped them to get licences for the Market on Wheels. Then the associations tried to exclude intermediaries from the market, and rallied priests and local teachers as enthusiastic supporters. What was more surprising was that Don Roberto—the American tomato agent—offered to provide contacts at the American border so that the Tlayacapan association could export directly. The association's representatives spent weeks obtaining export permits, and finally chartered two lorries for the first venture. But the border bureaucracy had to be bribed and the buyers turned out to be representatives of Dixon (Don Roberto's company). In the end, after discounting transport expenses, taxes and bribes, the payoff was not substantially different from that received by selling to Don Roberto—and the enterprise had wasted a great deal of time. It was not repeated.

Another undertaking of local associations has been to try to improve conditions at La Merced—but there is little that can be done. To threaten wholesalers by withholding tomatoes harms the farmers, not the wholesalers—who receive thousands of tons of tomatoes from other parts of the country (Sinaloa and Guanajuato in particular). The bulk of tomato production in Mexico is organised by big farmers, with large capital and technological resources; in addition, they receive preferential treatment from import–export companies. They would not consider making pacts with small-scale growers so that the latter could improve their market chances. All big-farm produce which is not exported goes to the market—permitting wholesalers to lower prices even below the level that market mechanisms normally dictate. Confronted with these facts the only thing the local associations can do is to develop the same strategies that groups of friends or kinsmen develop anyway—and more inefficiently, as conditions of trust do not always exist. They have also tried to get regular information on day-to-day prices. In fact, several daily newspapers provide price lists for the main urban markets (*El Día*, for Mexico City markets); but they report the selling prices to consumers, not the buying prices to producers. The difference between the two is sometimes striking.

The marketing of sugar cane and rice. The growing of both sugar cane and rice incurs low risks from weather variations, because lowland fields are less exposed to heavy storms and winds. Market fluctuations are not a source of risk either, as the prices of these products are subject to government control. In addition, farmers do not have to worry about marketing operations, as the harvests are automatically sold to the agencies financing the crops.

So far so good. But there is a risk factor in growing rice: the yield may decrease drastically if the fields do not receive enough water; and this happens if water is needed for sugar-cane fields. Most of the irrigation system is—it should be remembered—directly controlled by the refineries, and, in case of need, they do not hesitate to cut off the water to other lowland crops. The incidence of this factor is not always predictable (that is why it is a risk): one could hypothesise that the larger the area covered by sugar cane in a given year the higher the chance of rice fields not receiving water; but there is another factor which may affect the yield of a spring or the level of the Yautepec river, and that is the amount of rainfall. Excessive rainfall may cause serious damage to rice beds and decrease productivity—and, of course, refineries are not to blame for this.

There are also marked differences in the expectations of gain from rice and sugar cane, arising out of differences in the market value of the products and of the nature of the transactions with the financing agencies. The selling of rice to the Banco Ejidal is carried out through local *ejido* officers, and *ejidatarios* can supervise the transaction, and ensure that it is done correctly. They see that the product is properly weighed, and check upon the official prices of unhusked rice. Thus, if there are no irrigation problems, a gross gain of about 10,600 *pesos* per hectare (eight tons paid at 1,350 *pesos* each) may be expected, and a net cash payoff which would amount to 7,500 *pesos* per hectare after discounting the loan, interest and taxes, may be predicted. (However, lack of water or flooding may decrease productivity to 3 or 2 tons per hectare—and occasionally wreck the crop altogether.)

In contrast, farmers lose control of the sugar-cane process from the moment the operation of cutting is taken over by the refinery. The refinery transports the product to its own premises,where it is weighed without the farmer being able to check the operation. Farmers in fact claim that they are regularly cheated by the refinery in the weighing. Further, they do not know exactly how much the refinery is going to discount from their gross earnings. An average expected yield of one hectare of sugar cane is 100 tons. (First-season fields produce more than that; but ratoon fields yield a lesser quantity.) Assuming the refinery recognises that amount, the gross gain per hectare would be about 7,000 *pesos*. From it, the refinery discounts initial loans, labourers' wages, 9 per cent for annual interest, taxes, transport expenses, payment of medical insurance, plus other alleged charges such as cost of water pumping and previous debts of the *ejido*. Again, the farmers feel cheated. In these circumstances, an optimistic forecast of net cash payoff per hectare may only amount to 2,000 *pesos*—even less than the net payoff expected from a hectare of maize.[28]

Secondary crops and fruit trees

Before discussing agricultural decision-making processes, I shall very briefly describe some of the characteristics of the five significant secondary crops and fruit trees cultivated in the region (see Table 6.6).

None of the secondary crops (onions, beans, chillies, sorghum, peanuts) really provides an attractive and viable alternative to the main crops. Onions have almost as high risks as tomatoes, but less pay off; in addition, they are only profitable when irrigated, and have therefore to compete with sugar cane and rice.[29] Sorghum was grown experimentally in several places, but did well only in hot-country *ejidos*: this limits its expansion. Chillies and peanuts have advantages (low cost, low labour inputs; they are less perishable, and can be eaten by the family), but they are not attractive from the point of view of cash returns. As for climbing beans, they are not an alternative to maize: the two of them grow

Table 6.6
The Morelos Highlands: characteristics of secondary crops (1971)

	Cash cost per hectare (pesos)	*Man-days of work per hectare*	*Average productivity per hectare (1965) (tons)*	*Expected net cash gain per hectare (pesos)*	*Risk (weather)*	*Risk and Uncertainty (market)*
Peanuts	200	50	1.8	2,000	Medium	Low
Climbing beans*	200	60	0.5	900	Low	Low
Chillies	700	150	3.0	3,500	Medium	Low
Onions†	4,000	200	9.0	6,000	High	High
Sorghum ‡	1,500	200	4.5	5,000	Low	Low

* Sown in same field with maize
† Irrigated
‡ Grows in hot country only

together. Beans are very much appreciated in local diets (they are in fact the main local protein source), so they have an easy market. What they really do is reinforce maize growing for which they provide an attractive supplement.

Other crops are sown sporadically: wheat, peas, squash, cotton, barley are some examples in recent years. None of them has been successful enough to persist or spread.

To complete the agricultural panorama of the Highlands, it is necessary to make some reference to fruit trees. As mentioned, most households grow them in the back yards for domestic consumption. Some fruits—plums, coffee, lemons, guavas—are also sold in local markets, or from the home, in small quantities. The gain is meagre, and only provides a supplement to daily cash needs. However, in the upper *pueblos* (Tlalnepantla, San Nicolás, Felipe Neri), there are large orchards (between 0.2 and one hectare; exceptionally, 2 or 3 hectares) specifically devoted to commercial fruit trees, particularly to avocadoes and peaches. These commercial orchards require heavy family-labour investments, paid labourers at harvest time (also at other periods in the larger orchards), and lately investment in fertilisers and insecticides. Insecticide has not completely prevented the spread of fruit diseases; because of these, and also because of market fluctuations—for peaches especially—several fruit growers have found themselves in heavy debt. However, fruit orchards seem to hold a certain attraction as an alternative in terms of time, energy and land use. But fruit plantations have in fact expanded very slowly. There are two reasons for this: (*a*) the best land for fruit growing is actually occupied by timber forest, and its clearance is forbidden; (*b*) land devoted to fruit trees does not yield cash returns until the trees are mature. Avocado trees, for instance, only produce four years after planting; peach trees take five years to mature. Given the scarcity of land and the pressing need for cash, farmers can only afford to convert cultivable land into fruit orchards very gradually.

On the other hand, fruit trees are *the* cash alternative for most people in the Tlalnepantla *municipio*. This fact—rather than conservatism—explains why tomato cultivation has not developed in this *municipio*.[30]

The logic of decision making

From the foregoing, several simple points can be made about the overall pattern of farming decisions concerning different crops. In the first place, inexpensively produced crops such as maize (beans, peanuts and chillies to a lesser extent), which also can be used in the home, persist, despite their comparatively low price, because of the advantages they present to farmers without access to cash. In addition, as they do not require intensive labour their cultivation enables a small farmer to earn money by paid work. On the other hand, crops such as tomatoes present many disadvantages—they are expensive and risky—but they are attractive to people wishing to maximise cash income. As the social context of the Highlands demands cash simply to survive, it is likely that many people are willing to assume the risk and grow at least some tomatoes. However, it is unlikely that the proportion of land devoted to risky crops will increase appreciably in the future. In fact, farmers wishing to maximise should not grow tomatoes as their sole crop, lest they incur the risk of losing all the money invested.

Finally, the use of irrigated land must be considered in relation to how the refineries control the supply of water. The production needs of the refineries determine the amount of land and water available for the growing of more profitable crops such as rice. Normally, the majority of irrigated fields are only available between sugar-cane growing cycles. However, in spite of their protests, many people reckon that sugar-cane cultivation has several advantages: it is secure, and provides an excuse for continuously borrowing money from the refinery and investing it in other crops or simply using it for household expenses. The refineries know that farmers often use the loans for purposes other than sugar-cane growing; nevertheless, they continue providing cash since indebtedness increases the farmers' dependency (Ingham, 1968: 65 ff.). At the same time, the farmer knows that without sugar cane the opportunities for work during the dry season would decrease drastically. The majority of Highlanders get involved in cane cutting to some extent; although the refineries hire a number of permanent cutters, additional help is always needed.[31] Moreover, the transport of cane to the refineries is often partly done by farmers who have lorries—and who get paid for it. A number of people of the Highlands villages have industrial jobs in the refineries, and a vested interest in the continuation of the crop on the lands of their kinsmen and neighbours—some of them are *ejidatarios* themselves.[32] I am not arguing that sugar cane is objectively the best of all possible crops on irrigated land—probably it is not—but certainly, for a variety of reasons, the cultivation of sugar cane receives the approbation of many different local people.[33]

One could say, then, that some crops are favoured for their intrinsic benefits (money, security), others for their extrinsic advantages. But, in considering the various intrinsic benefits, is it possible to determine which benefit would be more likely to attract one type of farmer rather than another? Are there identifiable circumstances or attributes associated with 'money maximising', 'conservative', or 'target-income' farmers? If one looks at the distribution of crops for the fifty-two farms included in the Tlayacapan sample, it seems to vary widely (De la Peña, 1980: Table 6.9). However, it is possible to correlate the percentage of tomatoes actually grown by a farmer with the size of his farm. Yet this is not a simple 'positive correlation' (the larger the farm, the higher the percentage), but one which emerges when the sample is broken down into four categories (Table 6.7). On the basis of interview material, the people can be grouped as follows:

Category A is made up of people who in local terms could be considered as 'big shots': they hold roughly between six and eighteen hectares. They grow a lower percentage of maize than any other category; but not a higher percentage of tomatoes than categories B and C. There are several reasons for this. Cancian has argued that 'the higher a person's rank the more disadvantageous to him a random change ... is apt to be', because 'it will probably be downward rather than upward' (1972: 137). This 'inhibiting effect of high rank' may apply in relation to undertaking a risk. But on the other hand people in Category A have the opportunity of diversifying into crops other than tomatoes and maize, simply because they have more land. For instance, they are practically the only ones who grow rice. Even maize and sugar cane, when sown in relatively large areas, become relevant sources of income—secure income, which may be further invested in expensive crops, or in capital equipment (tractors, lorries), or in buying and/or renting additional land, without incurring debts or making great sacrifices.

Farmers in *categories B and C* tend to be more daring than farmers of category A (category C especially). Indeed, they have to be; because if they concentrated on low-risk crops, it is probable that the money thus earned would be insufficient even to make ends meet. In addition, these farmers are usually those who are more worried about their children's future. They know that their farms will be divided among their sons; they seek to buy more land lest the sons be reduced to a lower category. Alternatively, they seek to provide their sons with better schooling in order that they may be able to earn a living in non-agricultural occupations—and education usually costs money. Thus, medium-size

Table 6.7
Distribution of crops according to size categories (in hectares)

Category	*A*	*B*	*C*	*D*	*Total*
Size of farms (range)	5.8-18.1	2.7-4.7	1-2.5	0.3-0.9	0.3-18.1
Number of farms	9	15	17	11	52
Total amount of land	76.4	51.9	32.9	6.7	167.9
Percentage of total land	100	100	100	100	100
Total amount of irrigated land	22.3	12.6	7	2	43.9
Percentage of total irrigated land	100	100	100	100	100
Maize					
Area	33.3	26.8	18.4	4.8	83.3
Percentage of total land	43.5	51.6	56	71.6	49.7
Irrigated area	–	3.9	2.5	1.1	7.5
Percentage of total irrigated land	–	31.2	35.7	55	17
Red and green tomato					
Area	20	12.2	9.7	0.4	42.3
Percentage of total land	26.5	23.6	29.5	6	25.2
Irrigated area	1.6	3	3	–	7.6
Percentage of total irrigated land	7.2	24	42.9	–	17.2
Sugar cane					
Area	12.3	3.7	1.5	0.4	17.9
Percentage of total land	16	7.1	4.5	6	10.6
Irrigated area	12.3	3.7	1.5	0.4	17.9
Percentage of total irrigated land	55.1	28.8	21.4	20	40.8
Rice					
Area	6.4	–	–	0.5	6.9
Percentage of total land	8.3	–	–	7.4	4.1
Irrigated area	6.4	–	–	0.5	6.9
Percentage of total irrigated land	28.7	–	–	25	15.7
Others					
Area	4.4	9.2	3.3	0.6	17.5
Percentage of total land	5.7	16.7	10	9	11.4
Irrigated area	2	2	–	–	4
Percentage of total irrigated land	9	16	–	–	9.5

Source: de la Peña, 1980, table 7.9.

farmers are more willing to take risks, and their access to cash and loans enables them to do so to a much larger extent than farmers in *category D*.

This last category of farmers has very little land; they devote most of their time to paid employment. Even if they are willing to take risks, they usually cannot get large sums of money for investment because of the small size of their plots, and because they frequently have to borrow small amounts of money even to meet everyday household requirements. On the other hand, to grow less than 0.4 or 0.5 hectares of maize necessarily means an increase in household expenditure—which is not very wise for people in extreme need of cash.

Factors such as age, availability of household labour, size of household (in terms of members demanding food), years of schooling, years of experience in tomato growing, have no significant influence on the farming behaviour of the sample taken as a whole (see De la Peña, 1980, Appendix VII), though interviews convinced me that individual farmers are in fact influenced by some of these factors. (An old, childless farmer is not interested in expanding his farm, but in subsistence. Having sons to help in the fields decreases labour investment costs and encourages tomato cultivation. Illiteracy increases uncertainty in relation to marketing.) Hence the importance of a particular factor must be assessed in the context of the various sets of social relationships the farmer is involved in at different stages in his career.

Analysis of farming careers

The elements of risk and uncertainty have been singled out by some writers as being crucial in the analysis of decision making in a sociological context (Wharton, 1971: 578). Moreover, some anthropologists, such as Long (1970: 144–5) and Cancian (1972: 7), have argued that, given the existence of uncertainty, purely economic factors such as capital or technology are insufficient as explanatory variables for behaviour—and, thus, questions such as whether or not peasants are economically optimising become meaningless, as they only focus on the strategic allocation of economic resources. Indeed, empirical research on modern capitalist firms shows that their strategies are far from being purely economic, as these are often directed to the elimination of uncertainty through, for instance, the development of relationships of trust with competitors and clients and the emphasis on short-term rather than long-term goals (Cyert and March, 1963, quoted by Long, 1970: 145).

In describing farming strategies, I have shown that relationships of trust are needed at several critical points in the agricultural process—in hiring labourers, in acquiring credit, and especially in organising marketing. It was in relation to the selling of tomatoes that trust was important in order to confront situations of uncertainty. However, it should be pointed out that risk and uncertainty are related not merely to the moment in which actual marketing occurs, but to farming as a life career. Not only may a farmer lose money at the end of the harvest; he may also lose control over his land—and this frequently happens. This type of risk is often met by undertaking alternative employment and/or non-agricultural enterprises—labour migration, for instance, has been crucial for the continuation and development of farms—or by participating simultaneously or sequentially in other activities (commerce, for instance).[34]

It follows that farming decisions are better understood diachronically, as moments in a whole life career: 'What looks, *post hoc*, like a well made strategic decision (e.g. the diversification or consolidation of one's assets), on closer inspection may be re-interpreted as the outcome of a set of existing relationships and conditions which, to all intents and purposes, had cancelled out other possibilities; and if this is the case, then we cannot easily talk about a range of (present) alternatives' (Long 1975: 3). The same can be said in positive terms: there are relationships and conditions which favour some choices rather than others (Leeds, 1965).

By analysing the farming careers of three brothers I attempt to show how different factors become relevant to their farming decisions at different points in their careers. I also hope to show that the patterns of local farming occur in the context of the farmer's dependence on, and manipulation of, both local and wider social links and institutions. The following questions seem relevant in this connection: to what extent does inheritance condition the setting up of a farming enterprise, is household labour a crucial help in farming operations, what types of resources are mobilised by involvement in ritual networks, political factions, or family alliances? The answers may lead to a better understanding not only of agricultural dynamics but also of the social context of the Morelos Highlands.

The Figueroa family

Trinidad Figueroa (1888–1952)—a Tlayacapan farmer—had four children: Arturo (b. 1920), Manuel (b. 1923), Ana (b. 1925), and Pablo (b. 1928). He divided his land between only two of his sons, Manuel and Pablo. Arturo, however, is also a farmer—and a prosperous one. I will describe the expansion and division of Trinidad's farm, and then the life careers of his sons.

Trinidad's farm: Expansion and division

In 1925 Trinidad was confirmed as the owner of a plot of 2.4 hectares situated to the north of Tlayacapan. This plot had belonged to his father. By 1938 his holding had increased to 4.4 hectares, with the addition of three plots of irrigated *ejido* land which had been granted to him and his children (0.7 hectares to Trinidad, 0.8 to Arturo, and 0.5 to Manuel). All the land was controlled by Trinidad, as both sons were 'family children'. Arturo married in 1940, but remained with his wife in his father's household; Trinidad gradually let him take charge of some farming decisions. Manuel married in 1941, and in 1943 he moved to another house in the neighbourhood and set up an independent farm with his small *ejido* plot plus 1.4 hectares given to him by his father.

Before Manuel received his inheritance, Arturo had taken a paid job as a part-time officer of the *Banco Ejidal* in the rice programme. It was considered that because he had a supplementary income, he did not need more land of his own. Also Trinidad wanted Arturo to continue working on his own farm now that he had good connections in the rice business. In 1945, Ana married an *ejidatario*, and went to live with him without any inheritance. Pablo, the youngest son, married a girl from the *barrio* of Santiago (the Figueroas were from El Rosario) in 1946. His father-in-law gave them a house in that *barrio* to live in, so Pablo too left his father's household. With his move, Pablo inherited one hectare of private land. Trinidad gave him this rather than the *ejido* land, because he wanted to continue growing rice with Arturo. Trinidad and Arturo were left with their irrigated *ejidos* (1.5 hectares altogether) and with Arturo's salary from the bank—which seemed sufficient at the time. In 1952 Trinidad died. His *ejido* plot was inherited by his wife Herminia and administered by Arturo, who lived with her. With Herminia's death in 1962, this plot was inherited by Pablo, who was at that time the only son with no *ejido* land of his own. Neither Arturo nor Manuel could legally inherit the *ejido*, as no person is—at least in theory—entitled to rights over more than one *ejido* plot.[35] Arturo was then left with his own 0.8 hectares of *ejido*, and inherited his mother's house.

Arturo's farming career

Arturo had left primary school in 1932, after finishing the fourth grade, to work on his father's land, where, apart from some chillies, beans and pumpkins for domestic consumption, maize was the main annual crop. In 1936 he became an *ejidatario*, though his plot was worked as part of the parental farm. In 1943, when the bank launched its programme for rice cultivation, the family started growing rice in their *ejidos*, and Arturo was appointed local officer for the town of Tlayacapan. He keeps the office to this day. His functions include keeping a record of those *ejidatarios* who are potential rice growers, collecting loans from the bank branch in Cuautla, and distributing them among the *ejidatarios*. After the harvest, he has to weigh the rice, contact the buyers, and pay the producers. His yearly salary is 3 per cent of the total value of sales, plus expenses. By local standards it is high. The job involves work only at the beginning and the end of the season, and it provides him with a secure annual income. In 1944 and 1945 he used part of his salary to buy new steel ploughs and a pair of mules for himself and his father.

Arturo was appointed to this job by the representative of the bank in Cuautla. He was recommended for it by the local *comisario ejidal*, who was a friend of his father, and by the municipal president, who said that Arturo could read and write, and that he had proved capable in responsible posts—he had already held various public offices. Thus Arturo's job in the rice programme from the age of twenty-three has to be understood within the broader context of an active political career, which made him widely known and trusted in Tlayacapan. In 1940, when he was twenty, the municipality made him *ayudante*, again because of his literacy—which at the time was still rare among adults—and also due to his father's friendship ties with the authorities. In the following year he became *juez menor* (assistant magistrate for civil matters). Later, in 1949, he was appointed *regidor* (treasurer) of the *municipio*. In addition, his network of relationships was widened and reinforced by his active involvement in *mayordomias*, in which he participated from his early youth; he reached the top of the hierarchy three times.

1940–41	*Ayudante municipal*
1941–42	*Juez menor*
1943–	Rice officer in the *ejido*
1944–45	First *mayordomo* of Our Lady of Guadalupe
1946–47	*Padrino* of the Santo Niño
1949–50	Regidor
1956–57	First *mayordomo* of the Santo Niño
1970–	Officer of the CONASUPO

Up to 1952, then, Arturo was a popular figure, but he was not very successful financially, as he shared household expenses with his parents and had a family of his own to support. From 1941 to 1952, six children were born to Arturo, of whom two died shortly after birth. In 1952, after the death of Arturo's father, one of his children became severely ill, and Arturo had to borrow money from a Cuautla moneylender to take him to a Mexico City clinic. It was to clear this debt that Arturo first went as a seasonal migrant to the U.S.A., taking advantage of the opportunities given to Mexican agricultural workers by both the national and the American governments. He then repeated the trip every year without neglecting his irrigated plots or his bank job. During his years of seasonal migration Arturo managed to keep part of the profit of his Tlayacapan resources, thanks to careful planning and with the help of his brother Pablo. He would work in the fields until September, and then leave the crop of rice (or sugar cane, if the refinery had asked him to grow it) in Pablo's care; he then would spend from September to December working in the United States and come back in time for the harvest and to fulfil his office duties for the bank. Pablo received half the profit from Arturo's fields.

In 1962, Arturo's mother died, and Pablo became the holder of their father's *ejido*. At the time, the government was restricting permits to cross the border. However, this coincided with the tomato boom in the area; Manuel—Arturo's brother—was already growing tomatoes and doing very well. So Arturo decided to stay in Tlayacapan and try his luck at the new crop. He used his savings from seasonal migration to rent one hectare of private land and to buy equipment. The harvest was successful and sold well; the following year Arturo rented two hectares, on which he grew maize and tomatoes. From 1962 on Arturo has not had to migrate and has been able to continue renting land—in 1971 he was renting 2.8 hectares of irrigated *ejido* plus 3.4 hectares of private land. He has twice—in 1967 and 1970—failed in the tomato harvest; but he has been able to compensate the loss from his earnings in other products and from his bank salary. In addition, he has been able to bypass middlemen, and his brother Manuel owns a lorry which they use for transportation.

Arturo never bought any land, as he reckoned that to buy one hectare would cost him more than to rent ten. He could not really afford to buy land until the end of the 1960s. In addition, none of his children need land at present as all the older ones have non-agricultural occupations. During the 1950s, Arturo had access to only a small portion of the *ejido* (which was further reduced when his mother died), and encouraged his children to continue their education and find occupations outside agriculture. Three of them, Laura (b. 1941), Emilio (b. 1946), and Alberta (b. 1954), got government grants to study in education colleges in Cuernavaca and Tlaxcala and became teachers; only Laura has returned to the household and teaches in Nepopoalco. The other two live in Veracruz and Tlaxcala respectively. Another son, Luis, became an industrial worker in Cuernavaca; he lives with Arturo and commutes daily to his job. (Arturo's daughter married and left the household in 1968, and the two younger sons are at the local primary school.) As a result, Arturo has never been able to use any labour from within the household; but on the other hand he has no obligation to give land to his older sons if they get married, and his household expenses have been reduced. In addition, the salaries of two of his children, Laura and Luis, provide extra cash for the household. At the end of the 1960s, Arturo's economic success was patent in the material improvements he had made to his house, which had a new tiled roof, a gas cooker, a television set, a record player and other items of bedroom and living-room furniture.

In 1970 Arturo gained yet another opportunity of acquiring cash when the government-owned CONASUPO (Comisión Nacional de Suministros Populares) opened a store in Tlayacapan to provide fertiliser at cheaper prices throughout the Highlands. He was appointed local officer. This job took half his time; but he nevertheless continued to farm using hired labour. In 1971 the local CONASUPO store failed to receive any supply of fertiliser, so Arturo's job was interrupted; but he used the occasion to start a splendid business in retailing fertiliser and chemicals privately. Most farmers of the region buy the chemicals for the following season in January or February, when they have some cash from crop sales and chemicals are in good supply. In 1971, however, many farmers failed to buy at the usual time, because the government agricultural officer in Cuautla had promised them low-interest loans—both in cash and in fertiliser—from the National Agrarian Bank. The loans never materialised; in May, when people tried to get chemicals in Cuautla, prices had risen by 20 per cent. Arturo saw the opportunity of using the contacts he had made with private and State fertiliser producers during his years with the Ejido Bank and CONASUPO. He started buying chemicals on credit directly from the factories in Mexico City, transporting them in his brother's lorry and in hired vehicles, and selling them in the region at a price much cheaper than in Cuautla, but high enough to make a profit. The operation continued throughout the whole season; in November, Arturo estimated his commercial gains to be 40,000 *pesos*. He is planning to buy some land from the profits 'just in case any of the children need land after all'.

Comments. There are three distinct periods in Arturo's life as a farmer; the first, from 1932 to 1952, was conditioned by his role as 'family child'. Practically all his resources were directed to maintaining and running the parental household and farm, and subsidising his brothers in setting up as farmers. During this period, Arturo could have chosen to move away from his father's household, to live independently on the income from his *ejido* plot and his bank job; but there were several factors which prevented him from doing so. At the beginning of his married life, Arturo felt no pressures to have exclusive control over land and money; in fact it was more convenient to have access to the profits of a joint farm. He also expected to inherit part of it and, as his father grew older, Arturo became farm manager. After Manuel and Pablo left, Arturo had the responsibility of staying with his parents, and also the right of inheriting the house. The joint expenditure in the household did not leave Arturo or his father any savings further to expand the farm, especially after its size had been reduced by Manuel's and Pablo's inheritance.

In the second period (1952–62), Arturo became the head of the household but his access to land was reduced; in addition, the growing number of children and his son's illness increased family expenditure and forced Arturo to seek another source of income through seasonal migration. His children did not help on the farm, either because they were too young or in school; but he was able to keep the farm functioning thanks to Pablo's collaboration and, as all his land was irrigated, he was able to cover all labour and technological investments by loans from the refinery and the bank. The combination of resources in this second period allowed Arturo after 1962 to start a third period of expansion. He lost control of half the *ejido* land (when Pablo inherited his father's plot), but was able to compensate this loss by his migration savings, which enabled him to rent extra land, and by seizing the opportunity of producing tomatoes using the new technology. However, he would have been unable to make money out of tomatoes if he had not had access to cash for technology and labour: the supply of cash came from his migration savings and bank salary. Later he did not have to give either land or substantial help to the children who left the household, as they already had the means to support themselves; the older children who have stayed in the household have non-agricultural incomes and provide extra cash for the family. Finally, the fertiliser business was the result of a stroke of luck and proof of Arturo's commercial ability; but the condititions were (*a*) his job in the CONASUPO which, like the the bank post, is related to Arturo's acquisition of strategic social ties; (*b*) his connections with retailers in Mexico City as a bank official; (*c*) his already established economic position, which permitted him to hire extra labourers, pay for transport, and dedicate time to the business. In other words, this last financial success, like the previous ones, cannot be explained purely in terms of Arturo's personal qualities. It is necessary to take into account his position in a field of relationships which transcend family organisation and community boundaries. These have provided him with access to crucial resources which can be converted to investment in farming. Similarly, the pattern of his farming decisions at present has to be understood within the context of his career and accumulation of resources.

Arturo's farming in 1971

At the end of 1970, Arturo could count on approximately 42,000 *pesos* for the following year's expenses, of which 8,000 came from his bank salary, 12,000 from his CONASUPO salary, 2,000 from sugar cane grown in his *ejido* plot, and about 20,000 from rice grown on three hectares of rented *ejido*. He had also grown tomatoes on one hectare of private rented land, but the crop failed. His farming plans for 1971 were as follows: (*a*) to use his *ejido* plot for a second year of sugar-cane production; (*b*) to rent 2.8 hectares of *ejido* land for 3,500 *pesos* to grow rice, (*c*) to rent 1.7 hectares of private land for 1,000 *pesos* to grow maize, (*d*) to rent 1.7 hectares of private land for 2,000 *pesos* to grow tomatoes.

Expenditures in renting land thus amounted to 6,500 *pesos*. Investment for rice and sugar cane was covered by the bank and the refinery; investment for tomatoes (labour and chemical products) was over 20,000 *pesos*, and for maize over 2,000. That totals about 30,000 *pesos* for land and cultivation expenses. Arturo still had 12,000 *pesos* left plus the salaries of his children (Laura, the teacher, and Luis, the factory worker: they earn over 3,000 *pesos* a month between the two of them) to cover household expenses and to invest in the sale of fertiliser—which had a quick payoff.

What is obvious in Arturo's pattern of choices is the importance of cash obtained from non-agricultural sources and from secure crops like sugar cane and rice in enabling him to expand and diversify his farming. The risk implied in his decision to grow 1.7 hectares of tomatoes— even when the previous year's tomato crop had a negative payoff—is balanced by the possibility of his using a larger amount of land for low-risk crops, by the potential support of household members, and again by his own non-agricultural earnings.

Manuel Figoeroa's farming career

There are similarities as well as marked differences in Arturo's and Manuel's careers. Manuel was born in 1923; he also left school at twelve and, in 1938, was granted half a hectare of irrigated *ejido*. His independent career started in 1943, when he left his father's household to cultivate rice and sugar cane on his *ejido*, and maize, beans and peanuts on the plot inherited from his father. Manuel had no children, and he has never had economic problems either. His ceremonial life was even more active than Arturo's, having more money to spare; he was first *mayordomo* of an image for four consecutive years, hoping to get the favour of the Virgin and have a child.

1946–50	First *mayordomo* of Our Lady of Transito.
1952–53	First *Autor* in the Carnaval.
1954–55	Second *Autor* in the Carnaval.
1954–55	First *Mayordomo* of the Santo Niño.
1955–56	*Padrino* of the Santo Niño.
1962–65	*Comisario ejidal.*
1965–68	Municipal president.

In 1954 he started going to the U.S.A. for the harvest, following the example of his brother Arturo. He was very impressed by the modern techniques and organisation he saw there; in 1958 he stopped migrating and used his savings to grow tomatoes; he was among the first in Tlayacapan who tried improved seed and fertiliser. It was the time of the tomato boom in Mexico, and Manuel did very well. From that time, his farming has been continually expanding. He has bought 3.2 hectares of private land, a new house, a tractor, and a lorry. In 1961 he opened a shop, retailing tinned food, cigarettes, sweets, sort drinks and beer. This business has grown, and in 1971 it made a profit of 800 to 1,000 *pesos* per month. The shop is run by his wife, so Manuel can still concentrate on working his 5.1 hectares of land (0.5 *ejido*, 1.4 inherited private land, 3.2 purchased private land), as well as other rented plots.

In 1965 two of his sister Ana's sons came to live with him (Ana's husband has a single *ejido* plot, and only one of his sons expects to inherit.) These two young men work in Manuel's fields; one of them also drives the lorry. With their help, Manuel has been able to make extra money by transporting other farmers' produce to market, and by occasionally hiring out the tractor.

In the 1960s Manuel also became an important political figure. He had a wide network of friends through his previous participation in *fiestas* and carnivals, and he was also known as a hard-working, successful agriculturalist. He was elected *comisario ejidal* from 1962–1965, and then municipal president for the next three-year period.

Comments. Until 1954 Manuel Figeroa's economic life was determined by two main factors related to the household cycle: his inheritance, which allowed him to start an independent life, and the absence of children, which freed him from pressures of family expenditure. He also had control over an *ejido* plot from the beginning of his independent farming life. After 1954, the main determinants of his success were access to money from migration, which allowed him to grow tomatoes without incurring debts, good tomato harvests, and diversification of his activities by opening a shop. As in Arturo's case, the acquisition of money outside farming seems to be crucial in understanding Manuel's career. Manuel's 'take-off' was faster than Arturo's; Manuel himself explains this in terms of the different composition—and expenditure—of their households. It is also important to note that Manuel stopped migrating four years before his brother, and also started growing tomatoes earlier, and that he was able to compensate for the absence of household labour by adopting his two nephews. Manuel's inheritance allowed him to grow tomatoes on his own land, whereas Arturo had to rent a plot for that purpose. Later, Manuel was able to generate enough money to both rent and buy more land, whereas Arturo's expansion has been limited to the renting of plots.

Manuel was very reticent about giving details about his own budget in 1970 and 1971. All I could determine was that in 1971 he was farming 8.6 hectares of land, distributed as shown in Table 6.8.

Table 6.8
Manuel Figueroa's farming in 1971

Product	*Hectares*	*Type of land*
Tomatoes	2	Own private land (seasonal)
Maize	2.4	Own private land (seasonal)
Maize	0.6	Rented private land (seasonal)
Sugar cane	0.5	Own *ejido* (irrigated)
Sugar cane	1.3	Rented *ejido* (irrigated)
Rice	1.8	Rented *ejido* (irrigated)

The relatively large percentage of land at present devoted to tomatoes seems to be related to the possibility of diversifying crops in a farm expanded by capital accumulation, which in turn is related to access to both non-agricultural and agricultural income throughout his career.

Pablo Figueroa's farming career

Pablo's economic life, in contrast to that of his brothers, has never been very successful. Immediately after his marriage in 1946, he inherited a house through his wife, and one hectare of land from his father, and lived quietly on that until 1952, when Arturo asked him to help in the *ejido*.

Pablo has never migrated. Before 1952, he did not have much household expenditure, and did not feel any pressure to seek extra income. From 1952 to 1962 he had access to extra income from administering his brother's farm; this committed him to staying in Tlayacapan. After 1962 he obtained control of his father's *ejido*. But as time went by, Pablo became less and less able to save money. His household expenditure grew as family increased: from 1947 to 1965 his wife bore him nine children. His farming practices have always been on the conservative side—he has used his *ejido* plot for sugar cane and rice, and his private land for maize and beans. It was not until 1967, when two of his sons were helping in the fields, that he decided to grow tomatoes, and then only 0.3 hectares. In 1968 he grew 0.5 hectares of tomatoes, and has continued growing that amount in subsequent years. In 1971 he rented 0.4 hectares of private land, expanding his farm for the first time.

Also in contrast to his brothers, Pablo's public life has been very limited. The only major ceremonial office he has held was in 1965, when he became first *mayordomo* of Santiago. He has never held any political office.

Four of Pablo's children have already left school without going beyond primary level, and all except one married daughter live in the household. It seems predictable that when his sons get married, Pablo's small farm will be divided up.

Comments. As in the case of Arturo and Manuel, the initial course of Pablo's economic life was largely determined by inheritance and the household cycle. However, the subsequent stages of his career appear to be much more limited by his own household demands than in his brothers' cases, since Pablo did not have access to additional sources of income. It also appears that Arturo's opportunity to make money from migration was achieved at Pablo's expense; i.e. a necessary, albeit unintended, consequence of Pablo's administration of his brother's *ejido* was that he was not able to migrate, or to save much money, or to expand his farming.

It is striking that Pablo has not received any help from his brothers in later years. Whereas Arturo and Manuel are often seen together and co-operate in marketing tomatoes, Pablo remains rather aloof from his brothers. It may be that Arturo resented losing control of his father's *ejido*, and thus his relationship with Pablo became less cordial. Pablo lives in the *barrio* of Santiago, at the other end of the town, whereas Manuel and Arturo live in the same street and participate in the ceremonial and informal networks which develop among neighbours. Still more important is the fact that Pablo has not asked for their help. Pablo has never been in a situation of crisis; he does not live affluently, but manages to survive and feed his family. Probably, if he needed extra money for contingent expenditures, he would ask and get his brothers to help him. But, in normal circumstances, to work in association with his brothers would mean becoming dependent on them.

Relevant variables in farming careers

The purpose of the above analysis is not to unravel all the objective circumstances which result in given courses of action, but rather to emphasise the cumulative effect of piecemeal decisions in the development of a career. These decisions are in turn affected by the context in which an individual is active. It is possible to isolate some of these 'contextual aspects' on the basis of the cases of the Figueroa brothers, and then examine their relevance for other farming careers. For this purpose, it is convenient to define thirteen measurable variables and to see to what extent they are present or absent in the life histories of different types of farmers. The variables are as follows:

I. Inheritance of a reasonable amount of seasonal land (at least one hectare).

II. Inheritance of a reasonable amount of irrigated land (at least 0.5 hectares).

III. Personal grant of a reasonable amount of *ejido* land (at least 0.5 hectares).

IV. Inheritance of a house.

V. Functional literacy: at least three years of schooling.

VI. Labour migration (of any type).

VII. Ritual offices: in *fiesta* or carnival organisation as *mayordomo*, *padrino* or *autor*.

VIII. Political offices: both in the municipal council and in agrarian committees.

IX. Additional occupation: whether the farmer receives a permanent additional income as shopkeeper, government employee, etc. (Agricultural wage labour is not included.)

X. Additional income from other members of the household with paid occupations: teachers, government employees, seamstresses, etc.

XI. Permanent expansion of the farm: whether money has been invested in buying land.

XII. Temporary expansion of land: whether money has been invested in renting land.

XIII. Improvement of household facilities. I have selected eleven items, and given them values according to the cash expenditure they require: installation of new roof (3 points), tiled floor (3 points), living-room furniture (3), television set (3), water deposit (2), gas or electric cooker (2), record player (2), radio (1), electric iron (1), electric liquidiser (1), sewing machine (1).

These variables are significantly distributed in the careers of the farmers of the Tlayacapan sample, if the sample is broken down into the four categories previously used for correlating size of farm and crop distribution (see Table 6.7). Table 6.9 shows the percentage of farmers in each category for whom variables I to XII apply, and the average

Table 6.9

Life careers of farmers: Access to resources and social relationships

	Category	*A* (*N=9*)	*B* (*N=15*)	*C* (N=*17*)	*D* (*N=11*)	*Total* (*N=52*)
I	Inheritance of seasonal land	77.77	53.33	52.94	9.09	48.07
II	Inheritance of irrigated land	22.22	26.66	52.94	27.27	34.61
III	*Ejido* personal grant	55.55	53.33	5.88	18.18	26.92
IV	Inheritance of house	56.66	66.66	64.70	54.54	63.46
V	Functional literacy	77.77	60.00	41.17	36.36	51.92
VI	Labour migration	77.77	13.33	5.88	9.09	21.15
VII	Ritual offices	66.66	53.33	70.58	27.27	55.76
VIII	Political offices	77.77	53.33	23.52	—	36.53
IX	Additional occupation	55.55	33.33	35.29	45.45	40.38
X	Income from children or dependants	33.33	13.33	29.41	—	19.23
XI	Permanent farm expansion	44.44	40.00	17.64	—	25.00
XII	Temporary farm expansion	55.55	20.00	17.64	9.09	23.07
XIII	Household improvements	11.00	6.5	6.8	4.63	7.05

N.B. Variables I to XII: percentages; Variable XIII: averages.

number of points in each category for variable XIII. It should be noted that access to land at the beginning of a career (variable I, and variables II and III taken together) appears to be crucial as a differentiating element—which emphasises the importance of local relationships and institutions; labour migration—which emphasises the importance of external links—is also a key factor (see the life history of Rolando Campos in Chapter V). Participation of household members—including the farmer himself—in non-agricultural activities (variables IX and X together) also seem relevant in a double sense: both as a support to undertaking farming on a comparatively larger scale (category A), and in providing a supplementary income (from pottery, for instance) for poor farming (category D).

A more precise evaluation of the impact of these variables and of their inter-relationships would need more sophisticated statistical techniques—or more detailed accounts of life histories; however, I believe that the relevance of the variables is now clear in a broad sense for understanding the different patterns of life careers, which are in turn related to different patterns of farming decisions. It is now possible to look at strategies of economic maximisation in a context which is much wider than the agricultural activity; i.e. agriculture is one among several methods of gaining increasing access to cash. From this perspective, it is understandable that the farmers of category A incur less farming risks than farmers in category C (the latter have to rely more heavily on agriculture).

A useful way of looking at the involvement of farmers in a variety of activities is in terms of Fredrik Barth's well-known 'entrepreneurial' model (1963). Barth defines entrepreneurial activities essentially in terms of innovation and expansion. He draws attention to three operational aspects of such activities: the actor's *niche* or position in respect to resources; his *assets* (capital, skills, social claims)—what he uses to exploit the niche; and the physical, interactional or institutional restrictions he might face. In addition, Barth suggests that these three aspects become manifest in the transactions through which an actor participates in one or more 'spheres of value'. For instance, there is a distinguishable sphere of value in agriculture, and it is only in this sphere that assets such as expertise in ploughing become relevant—such expertise is not valued in the sphere of commerce, for example. Innovation, according to Barth, takes place when an actor is able to transfer resources and assets from one sphere to another. Expansion without innovation occurs when profits are converted into assets within the same sphere. Both innovation and expansion are found in Tlayacapan: from migration cash to farming land, from prestige in politics to permanent jobs, from trust in kinship or ritual networks to marketing. It is interesting that in Table 6.9 farmers of category C appear especially active in ritual, and especially interested in household improvement (they score higher than farmers of category B on variable XIII). Both household improvements and ritual organisation are important elements in the acquisition of social prestige—which facilitates the development of friendship and trust. Lacking other assets, this type of farmer attempts to manipulate and convert all possible local social links into economic value.

A final point must be made about the type of enterprise which emerges in the context of the Morelos Highlands. It is not a large-scale enterprise which monopolises land and 'proletarianises' its labourer. (Farm expansion is as likely to occur by renting as by buying land. Manuel Figueroa's investment in a shop, and Arturo's in his children's education are representative of a common strategy of transferring agricultural profits to non-agricultural spheres.) Nor is it a Chayanovian household imprisoned within an equilibrated production–consumption circle. It is a flexible entity confronting a difficult environment by manipulating multiple assets and resources. It is not isolated or independent of the influence of *neolatifundium* types of enterprise which operate in the wider society: refineries, technology peddlers (sellers of seeds and chemicals), wholesalers. It is also influenced by the action of government agencies: the bank, CONASUPO, the agrarian delegate, and the Ministry of Hydraulic Resources. And it is interlocked with enterprises of a similar kind: farmers' careers are interdependent; enterprises compete as well as co-operate with one another; the expansion of one farm occurs at the expense of the contraction of another.

Notes

1. The concept of 'achievement motivation' was firstly developed by McClelland (1961), who in turn draws upon Max Weber's concepts on the impact of the Protestant ethic.

2. Godelier's distinction between 'rationality of system' and 'rationality of actor' might be pertinent in this connection (1968: 17–19). According to Godelier, the system determines the priorities of action; actors are rational if their behaviour is consistent with the pursuit of certain explicit goals; but these goals are not created by the actors but by the system.

3. During the 1960s, the expansion of the cultivated area in the whole of Morelos reached its maximum possible limit (Garza-Sánchez *et al.*, 1968: 18).

4. There has been a slight increase in average productivity per hectare after 1965, for maize (1·2 tons), and for red tomatoes (14 tons), but not for rice (unpublished data provided by the Centro de Inversión y Asignación de Recursos, Consejo Nacional de Ciencia y Tecnología, Mexico City, 1973).

5. There is one inconsistency in the census data which should be at least tentatively explained: put together, the sugar cane and rice areas in 1965 cover a number of hectares which is larger than the total space of irrigated land supposedly existing in Tlayacapan in 1970. This may arise out of the fact that water is often cut from plots which are not currently devoted to sugar cane or rice—i.e. the amount of irrigated land may vary from one year to the next.

6. See, for instance, the statements made by the Minister of Agriculture to the newspaper *El Día*, 2 October 1971.

7. My personal impression is that in most parts of the Highlands maize covers about 70 per cent of the fields.

8. A farmer would need at least fifty hectares of land in order to use the tractor at full capacity. Thus, rather than as a straight-forward investment in individual farming, the acquisition of a tractor is regarded as a source of extra money as the tractor can be hired to others.

9. There are two varieties of sugar cane which are grown in the region: 'White Castilian' and 'Purple Veracruz'. The latter is usually preferred by the refineries. Sugar cane fields are burnt to get rid of the abundant, hard-edged leaves, which may hurt the hands of the cutters.

10. Chazanov rightly remarked that the economics of small-scale rural production are not explainable by the rules of capitalist accounting.

11. Hired labourers, though, frequently provide their own hoes and machetes.

12. The use of tomato seeds from a previous harvest produces very low yields, which do not compensate for the heavy investment in fertilisers and fungicides.

13. The poles used to be cut from hill bushes (usually from a local bush called *cuilote negro*); but nowadays, after ten years of bush devastation, they have to be bought in Cuautla, and replaced every three or four years.

14. Legally, the minimum salary in rural Morelos is twenty-four *pesos* a day; but usually *oaxacos* are paid fifteen *pesos* or less because they are allegedly unskilled and slow. Local people accept wages of twenty *pesos* a day. As the labour force on most farms consists of both migrant *oaxacos* and local labourers, I have estimated the average cost per man-day at eighteen *pesos*. A report of the Ministry of Hydraulic Resources which I saw in the municipal offices of Tlayacapan estimated total costs of red tomato production per hectare at 4,096 *pesos*. My data suggest that this is a gross underestimate.

15. The *ingenio* of Oacalco finances growing of sugar cane in the *ejidos* of the town of Tlayacapan and of the villages of Pantitlan and San José de los Laureles. The *ingenio* of Casasano operates in the *ejidos* of San Sebastian and San Andres.

16. All government agricultural banks merged into one single institution at the national level in 1973. However, many local agencies still operate independently.

17. It should be remembered that a good many farmers do not have properly legalised deeds.

18. The farmers also claim that the fertiliser is charged by the banks at a price higher than the ordinary retail price.

19. There is a local association in Tlalnepantla which has managed to continue receiving loans from the Banco Agropecuario for tomatoes. But its membership is restricted to twenty people only.

20. The loans from the banks and the refineries have an interest of less than one per cent per month.

21. The local *cuartillo* is equivalent to two litres.

22. A federal government agency—the Confederación Nacional de Suministros Populares (CONASUPO)—built a maize store for the Morelos Highlands in 1967 near Tlayacapan. It does not receive much maize, because CONASUPO insists on buying by weight and not by volume. The most popular local variety of maize (*anchoblanco*: it tastes much better, according to local opinion) is voluminous and light, so people prefer to sell it elsewhere. In addition, the CONASUPO often finds excuses not to pay the full price: humidity, grain deformation, etc. The hybrid variety of maize which is the heaviest of all is considered not to be good enough for local food standards.

23. The Banco Nacional de Crédito Agrícola recently opened facilities for insuring crops on smallholdings. The cost of insurance is 150 *pesos* per hectare; the compensation received in case of failure, 600 *pesos* per hectare. This compensation is obviously insignificant in relation to investment in tomatoes, but it is used for maize, which provides still another element of security for the cultivation of maize.

24. 'Risk describes situations where an actor knows the odds for and against a desirable outcome from a given course of action; uncertainty describes situations where he does not' (Cancian, 1972: 6–7). From the point of view of the actor, uncertainty is fundamentally caused by lack of information.

25. The nearby market towns of Ozumba and Jojutla provide some demand—never enough for all the production.

26. This agent (Don Roberto) has a local partner (the manager of the bus station and the telephone booth in Tlayacapan). Recently, Don Roberto (a Texan Mexican-American) married the daughter of his partner. Thanks to this connection, he is able to get information on where the best tomatoes are being harvested, and what price the other middlemen are offering. He then bids a slightly higher price.

27. 'Cognitive dissonance' is a term coined by Festinger *et al.* (1964) to refer to contradictions between two beliefs or items of knowledge. They argue that dissonance tends to be eliminated from perceptions about actual behaviour as long as circumstances permit; i.e. a belief which contradicts a course of action which has been already taken is usually put aside.

28. A new Sugar Law (1975) demands sugar cane to be paid according to the amount of saccharine it contains; however, the techniques for measuring this are still extremely faulty.

29. Some people in Atlatlahucan have begun to grow onions in back yards, or in fields very near the town, and irrigate them with buckets of water brought from the *jagüey*. For obvious reasons, this method can only be used for small fields.

30. See de la Peña, 1980, Tables 6.1–3. In 1971 the mayor of Tlalnepantla estimated that 200 hectares in the *municipio* were covered by tomatoes.

31. Wages for cane cutting are practically twice the ordinary agricultural wage. In 1971, the payment was 6.5 *pesos* per ton cut. Working hard, a man can cut from six to

eight tons in a single day.

32. Any refinery has two types of industrial workers. One works the whole year round; the other—more numerous—only during the six months of the *zafra* (December to May). This second category often farms as well.

33. Maize grown in irrigated fields produces two yearly crops. Alternatively, a crop of maize in the rainy season can be followed by a crop of tomatoes in the dry period. An irrigated field of tomatoes in the dry season presents less expense and weather risks, and it has much better chances of yielding higher cash profit, for market prices tend to rise at that time of the year as the national supply is lower.

34. The existence of 'multiple enterprises' whereby individuals get involved in more than one branch of economic activity has been described as a major characteristic of Third World economies, where resources are scarce and fragmented. It has also been argued that the analysis of the processes from which such enterprises emerge and develop is essential for understanding the social reality which conditions their existence (Long, 1975).

35. They could have tried bribery; but Pablo would have opposed this and a family fight would have developed. Besides, it was Herminia's will which decided the matter.

VII
CONTEMPORARY RITUAL DIMENSIONS

Life is slight as a thread. We exist today and disappear tomorrow. If life is only a loan, we should join together and help each other [Librado Sánchez, Tlayacapan farmer]

In earlier chapters, I criticised those conceptions of Mexican village life which present it as embedded in a timeless, untouched social and cultural structure. Implicitly or explicitly, such conceptions correspondingly view national society as having a 'plural' or 'dual' character, composed of segments with different economic and cultural institutions.[1] The model of plural society has been applied mainly to colonial situations; but some scholars have argued that, in the case of Mexico, a regime of 'internal colonialism' persists (González Casanova, 1965). It is argued that there is a relationship of dominance between the urban-based elite and the rural 'Indian' population, by which the latter is kept as a reservoir of cheap labour and denied participation in the dominant culture. The 'Indian' community thus tends to become an encapsulated entity: encapsulation is the consequence of a process of self-defence against exploitation and pillage (Frank, 1971: 149 ff.; Aguirre-Beltrán, 1967). The solidarity of fellow-villagers is manifested in their commitment to intra-community transactions, which in turn finds a symbolic expression in the organisation of communal ceremonies.

This model, as was the case with some of the conceptions of peasantry, by stressing cultural separation and single-stranded relationships of dominance, ignores the facts that both in the past and in the present (*a*) the members of different 'segments' have and had common as well as opposed interests in many fields of social activity; (*b*) the members of a single 'segment' have been and are divided by heterogeneous interests; (*c*) the proliferation of bonds and alliances among members of different segments has sometimes outweighed the importance of the commitments within a cultural or local group.[2] In addition, I find the term 'Indian' of dubious utility; even as a contemporary folk concept its meaning is unclear (Pitt-Rivers, 1965). The Morelos Highlanders apply the word to *oaxaco* immigrants and they were shocked when I asked if it could be applied to themselves. But to urbanites of Mexico City or Cuernavaca, the Highlanders, or at any rate some of them, are 'Indian'.[3] Perhaps because of the rejection of 'Indianness' the Nahuatl language is rapidly disappearing in the region: it is commonly spoken only by old people in the *pueblos* of Tlalnepantla, Totolapan and San José de los Laureles.[4] In Morelos—as has been shown by Judith Friedlander's phenomenological analysis—the words *indio* and *indígena* have purely negative connotations if applied to living persons.

It is also inadequate to look at local ceremonies as institutions which are culturally separate from the larger society and which function in terms of 'the maintenance of an already established sociocultural order' (Cámara, 1952: 146). Such an approach does not apply to colonial *fiestas*; neither does it apply to present-day celebrations. These events have many different functions and are open to several levels of meaning according to the position and interests of the actors involved.

An examination of the various levels of meaning may throw further light on issues such as the organisation of agricultural labour, the functions of the household, and the importance of local commitments and trust relationships. In the following chapter, I compare the standpoints of local actors with those of church representatives and the social relationships which emerge in ritual with those which emerge in formal politics. Both comparisons are important for understanding the connections between local ritual and the wider society.

The territorial framework of Tlayacapan fiestas

The two main streets in Tlayacapan form a cross oriented to the four cardinal points. At the centre are the town square and the main church; four chapels mark the extremities of the cross. These are Santa Ana, El Rosario, Santiago and Exaltación, and each gives its name to the quarter or *campan* in which it stands. The four *campans* of the town are roughly triangular in shape; their areas can be visualised by imagining a second cross superimposed diagonally on that formed by the main streets. The central square, the main church and the several ravines which run through the town serve as boundaries to the four quarters.[5] Around the central square is an area which does not clearly belong to any of the quarters; the inhabitants are sometimes referred to as 'the central ones'. In the past the 'non-Indian' elite lived there and the area is still favoured by some of the local well-to-do.

When a Tlayacapaner is asked where he lives he usually gives the name of his *campan*. Some people use the word *barrio* to refer to a *campan*; but *barrio* is also used to refer to a group of people who organise and/or co-operate in the celebration of *fiestas* or religious events in honour of the saints of a particular chapel. For the sake of clarity I shall use the word *barrio* in this latter sense. Each *campan* has several active chapels and hence several *barrios*. Each chapel may house more than one saint, and then it is said to have several *mayordomías*. A *mayordomía* is the organisation responsible for the *fiesta* of a particular saint and has its own hierarchical structure.

Table 7.1
Tlayacapan sections: households and chapels

Section	*Total number of blocks*	*Number of empty blocks*	*Number of households*	*Number of active chapels*	*Number of abandoned chapels*
Santiago	45	24	122	2	5
Exaltación	16	3	42	1	1
Rosario	20	7	118	4	2
Santa Ana	23	9	187	3	2
'Central'	8	–	44	–	–
Total	112	43	513	10	10

Source: Census for Water Taxes (1971)

Neighbourhood and ritual. Not all parts of the town are equally densely populated: only 40 per cent of the blocks have inhabited houses (Table 7.1). Figure 2 shows that most households (represented by small circles) cluster around the two main streets in the vicinity of the four terminal chapels. People's awareness of belonging to a particular quarter stems from the fact that they live in a neighbourhood in which people are in frequent contact with one another. During the day, while men are away working in the fields and children are in school, the women take

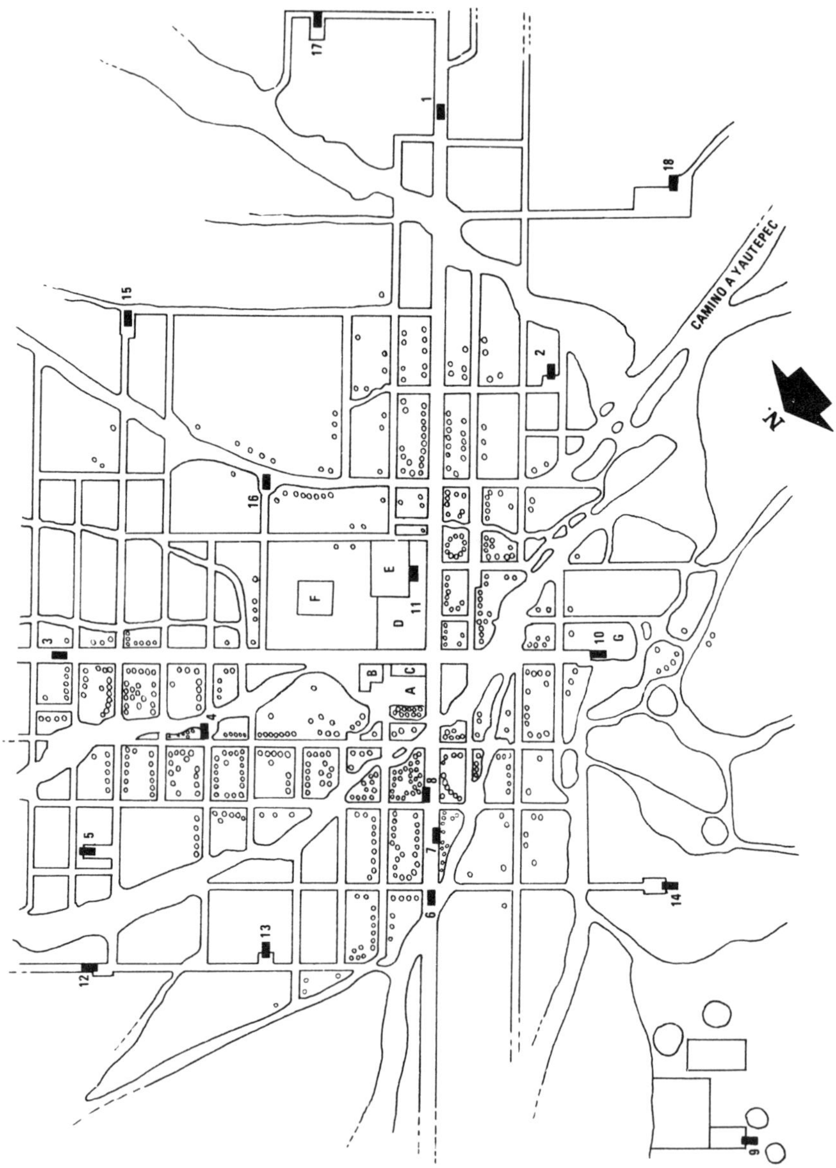

Figure 2. The layout of Tlayacapan. *Key:* A Central square, B Municipal palace, C Platform for open market, D Churchyard, E Augustinian monastery, F Primary school/Well, G Graveyard. 1 Santiago, 2 La Concepción, 3 Santa Ana, 4 Altica, 5 Magdalena, 6 El Rosario, 7 San Martín, 8 Tlaxcalchica, 9 La Villa, 10 La Exaltación, 11 Main church, 12 San Jerónimo, 13 San Mateo, 14 San Lorenzo, 15 San Miguel, 16 San Nicolás, 17 La Natividad, 18 Los Reyes.

care of the smaller children and prepare the evening meal; they also visit their neighbours, exchange gossip and borrow small quantities of food. In the evening, when the men are back from work, the streets of the neighbourhood are the scene of open-air informal gatherings; the women bring chairs out of the houses and talk; the children play in groups, and the men gather on the street corners to talk, play the guitar and drink. In these informal meetings, there is no differentiation between farmers and labourers, poor and less poor; in fact, all four neighbourhoods have people of all economic statuses. It is not that members of different *barrios* do not meet: they do, in the local market and in the bars of the central square; but the point I want to stress is that most informal encounters occur among neighbours. A fourth of the Tlayacapan marriages during the last twenty years took place between members of the same *campan*; for the last fifty years the proportion is 34 per cent of all marriages, which means that there are many blood and affinal kinship relationships among neighbours (see Table 7.2).

One consequence of this pattern of 'multiplex' relationships is the difficulty of clearly defining rights, obligations and rank in everyday encounters. 'Friends' and 'kinsmen' are words which can be applied to most neighbours. And one expects help and co-operation from kinsmen and friends. Yet social reality imposes limits to co-operation: resources are scarce; local people have to compete with one another for land, jobs, loans. There is conflict between open conviviality and the limits to generosity imposed by scarcity. 'To be *confianzudo* (too much confident)', 'to be *gorrón* (scrounger)', 'to be disrespectful' are

Table 7.2
Ratios of marriages in Tlayacapan

	1	*2*	*3*
Before 1930 (N=80)	0.43	0.37	0.18
1930–50 (N=46)	0.23	0.50	0.26
1950–71 (N=35)	0.25	0.57	0.17
Total (N=161)	0.34	0.45	0.20

1. Ratio of marriages betwen members of the same *campans*
2. Ratio of marriages between members of different *campans*
3. Ratio of marriages to outsiders
(Ratio = $^{n}/N$)

Source: Tlayacapan sample

reprobatory expressions frequently used in conversation. But equally condemned is he who locks himself up in his house and avoids his neighbours: expressions such as 'egoism' or 'vanity' are applied to that sort of behaviour. If a man gives a birthday party, all neighbours are invited. The door of his home remains open all evening. But only those who are likely to reciprocate actually attend. Furthermore, most of the people who attend the party are likely to have a more precise link—what we may call a ritual link—with the host than that of 'friend' or 'neighbour'. (This rule excepts youths and girls, especially if the occasion is a dance.)

A balance between 'egoism' and 'scrounging' is sought by establishing precise conventions based on a consensus about the appropriate degrees of distance and proximity. It has been argued that ritualised relationships achieve this balance, in addition to providing a means of defining alliances within an amorphous, unbounded social universe (Gluckman, 1955; Bonfil, 1973: 243 ff.). I have already suggested that *compadrazgo* links provide a way of reinforcing co-operative ties between particular individuals within a wide set of kinsmen. The ritualised (i.e. repetitive, and sanctioned by reference to a supra-worldly code)[6] behaviour of *compadres* is a continuous reminder of their peculiar link: they speak to each other with the term *usted* (the respectful form of address in Spanish) even if they have been close friends and drinking companions before becoming *compadres*. They stop in the street when they meet, shake hands, and inquire about the welfare of the other's family. When meeting a baptism, confirmation or wedding godfather (or godmother), the godson is expected to kiss his godfather's hand. *Compadres* visit each other on their respective birthdays and namedays. If there is a party, they are careful to arrive before anyone else. They go into the room where the domestic altar is—each household has a special place where the saints' images and frequently the family photographs are displayed, usually with candles in front of them—and presents are given to the host: a bag of bread and sweets, a handkerchief, sometimes a piece of cloth (but this imposes too much of an obligation: a small gift is more appreciated); finally, they embrace each other and share a cup of hot cocoa. Young godsons are sent by their parents to visit their godparents 'so that they receive advice' (*para que los aconsejen*). A quarrel with a *compadre* or a *padrino* is deemed to be a sin: 'almost worse than a quarrel with your parents'. In *rites de passage*, *compadres* and *padrinos* play a crucial role (Ingham, 1968: 144 ff.). A girl is requested for marriage by both the father and the baptism godfather of the groom. The bride is taken to church by her godfather. The groom's wedding godfather throws a big party, which is called *velación*, the night before the ceremony, in his own house. When a man dies his *compadres* keep company with the widow and contribute to the funeral.[7]

Fiestas and the differentiation of relationships. The basic definition of a *fiesta* is that it is a celebration organised by a set of neighbours, for enjoyment and as an expression of religious feelings, in honour of the saints of the neighbourhood. Thus a *fiesta* serves to identify members of the *barrio*. This category can be subdivided into *mayordomos* (those responsible for organising the *fiesta*) and the rest. This differentiation reflects differences in economic status. *Mayordomos* have to pay for a substantial part of the *fiesta*, usually between 400 and 600 *pesos* each. As a labourer can only earn 600 *pesos* by a month's hard work, he is unlikely to become a *mayordomo*. However, he may contribute a small sum to the *fiesta*, usually 20 or 30 *pesos*. The same applies to small farmers. However, for a farmer whose yearly income is 20,000 or 30,000 *pesos* the expenditure of 400 or 600 *pesos* does not imply a redistribution of his capital resources, nor is it a display of wealth;[8] but it is a way of publicly defining his superior status *vis-à-vis* his neighbours with whom he has multiplex relationships. In this respect, *fiestas* contrast with other encounters between neighbours, where differences in status are obliterated by a web of balanced exchanges of conversation, gossip and recreation at an informal level. Of course, status differentiation also emerges in the formal work relationships between a farmer and his labourers; however, such situations involve only a few people, given the small size of the farms. In addition, the social units in work relationships are discrete, whereas *fiestas* involve a large number of people and focus their attention on the performance of ritual activities in which the actors are hierarchically classified.

This level of definition of the *fiesta* is confirmed if we consider those people in the town who never take part in *fiesta* celebrations. These are: (*a*) the fourteen teachers at the primary school; (*b*) the owners of the two big stores; (*c*) the manager of the bus station (who is also a restaurateur, a representative of the telephone company, and the local agent for the Dixon To-ma-Toe Company); and (*d*) a group of five families who live in the area between the *barrios* of Santiago and Santa Ana, and are not clearly members of either.

The teachers present a public image of being aloof from religion; they never criticise the *fiestas* in public, but in private conversation they say that the cult of the saints is mere superstition. The storekeepers and the bus manager, on the other hand, do not hide their contempt for the *fiestas*, which they consider to be a waste of money and a pretext for the men to get drunk.[9] They also hint that *mayordomos* steal part of the money they collect. Both the teachers and the store owners do not need to resort to *fiestas* to display their status publicly; both perform roles which are socially well defined.

The five families who live between the *barrios* of Santa Ana and Santiago not only remain aloof from *fiestas*, but also rarely mix with anybody else. They are in fact converts to Protestantism. Their houses stand in the midst of several empty blocks, and their occupations do not involve much contact with other local people. Three families run a poultry farm, and they sell eggs and chickens to urban middlemen; another family head trains dogs for week-end hunters who come from Mexico City to explore the mountains of the region; the head of the fifth family is a photographer who spends most of his time working in the lowland towns.

The sequence of fiestas. The annual cycle of a *fiesta* reveals a number of different functions carried out by groups of participants, whose status is publicly recognised. The *fiesta* organisation itself serves to distinguish between *mayordomos* and non-*mayordomos*, and the former can be split into two categories: the first (chief) *mayordomo*, who is responsible for the overall organisation, and the secondary *mayordomos*, who assist him. The annual cycle begins when the first *mayordomo* is chosen for the coming ritual year. This takes place at end of the current cycle. A man who wishes to be the first *mayordomo* of a saint has to apply directly to the current first *mayordomo*. The position is granted to the first man who asks for it, provided he is considered to be a member of the neighbourhood; in other words, he must live in the *campan*. He does not necessarily have to live close to the chapel of the saint. The petition for office is not marked by a special ceremony, but it usually takes place before witnesses to avoid misunderstandings or personal preferences.[10] However, the formal appointment to the office is made in the presence of all the *fiesta* participants, including the former first *mayordomos*. This ceremony highlights the status of both old and new incumbents. The new first *mayordomo* has than to appoint the secondary *mayordomos* for the *fiesta* and he will choose them from among his personal friends. (Their number can vary from year to year and from *fiesta* to *fiesta*, but it does not usually exceed six or seven.) Thus, the main organiser is given the opportunity of defining more precise links with a chosen set of neighbours.

The next ceremonial step in the cycle—which is called *recibimiento*—is the removal of a glass casket, which contains a miniature replica of the image of the saint, from the house of the former first *mayordomo* to the house of the new one. The casket is carried in procession by the two first *mayordomos*—the old and the new—followed by the secondary *mayordomos* and their wives. By this procession, the new secondary *mayordomos* are presented to the neighbourhood. Frequently, the *mayordomo* invites all neighbours to ice cream and biscuits (*soletas*). The casket is then put on a specially prepared altar in the front room of the first *mayordomo's* house, so throughout the year all visitors and bypassers are reminded of the ritual status of the inhabitants of that house. This altar is permanently dressed with fresh flowers and candles brought by the *mayordomos'* wives. In addition to the casket, the first *mayordomo* receives the key to the chapel where the principal image of the saint is lodged. Throughout the year, the *mayordomos* and their wives are in charge of keeping the chapel clean, renewing the flowers and candles in front of the image, and making sure that nothing is stolen.

Nine days before the day of the saint—marked by the Catholic calendar—a ritual period known as *novenario*, or *novena*, begins. The casket with the small replica of the image is taken in procession from the house of the first *mayordomo* to the chapel. The procession is announced by a burst of rockets fired by the *mayordomos*; then it proceeds to the chapel, accompanied by a band of musicians and many people from the neighbourhood. The streets are adorned with flowers and ribbons made of coloured paper. When the procession arrives at the chapel the rosary is recited. Only the women participate in the rosary; the men stand outside the chapel during the prayers. The rosary is led by a group of women known as ritual specialists. Each neighbourhood has its own group. They know the litanies of the Virgin Mary in Latin and lead the hymns at intervals between the prayers. They are formally appointed by the wife of the first *mayordomo*. The rosary is repeated every evening during the *novena* period; invariably it is attended only by women.

At the end of the *novena* the *fiesta* itself takes place. It consists of the mass and a public banquet both of which are attended by many people. The *mayordomos* and their wives are responsible for the preparations and each has a specific set of tasks. The *mayordomos* are in charge of hiring the priest and the musicians, ringing the bells, firing rockets and preparing alcoholic beverages; the wives clean and decorate the chapel and prepare the food.

The *mayordomos* share most of the cost of buying flowers, candles, rockets and drink for the whole cycle; but small contributions (20 to 30 *pesos*) are also made by 'companions' (*compañeros*)—friends, relatives and immediate neighbours of the *mayordomos* and other members of the neighbourhood. The *mayordomos'* wives have female 'companions' who contribute some food for the banquet.

The banquet is attended by a wide range of people: the *mayordomos* and their wives; those who have held the office in previous years; the ritual specialists and their husbands; the 'companions'; the musicians; friends and relatives of the *mayordomos*, including people from other *barrios*; and people living outside Tlayacapan but with close kin in the *barrio*. The presence of the latter is highlighted by shouts ('Look who is here! Ah, you have not forgotten your little *pueblo*!' and similar expressions), hugs and toasts. Customarily, migrants give small sums of money to the first *mayordomo*; if a migrant is appointed secondary *mayordomo*, then his contribution will be more generous. (Rolando Campos, for instance, was a secondary *mayordomo* for several years while he was a migrant; see chapter V.)

At the end of the banquet, a new first *mayordomo* is appointed, and a new ritual cycle begins.

Campans, barrios, towns. Carnival feasts are organised by the individual *campans*. Like the *fiestas*, they provide occasions for differences in status to be displayed; there is a distinction between the sponsors—*autores*—and the rest of the people, and between the first and the secondary *autores*. In the past, the 'central' people (those living around the central square) organised a separate carnival, supposedly displaying more wealth and elegance. Nowadays, this no longer takes place. Exaltación and Rosario used to organise the carnival separately; now, they do it together. The reason given for the merging is that the population in both sections has decreased. (This contention is puzzling: the population of Tlayacapan decreased from 1910 to 1930, but rose thereafter. But people argue that, for some reason or other, the increase took place mainly in the sections of Santa Ana and Santiago.)

The organisation of the carnival cuts across *mayordomía* affiliation and thus provides a common framework of social differentiation for people from different *barrios*. Similarly, the *fiestas* of the saints lodged in

Table 7.3
Tlayacapan: chapels, images and celebrations (1971)

Chapel and image	*Type of celebration*	*Date*	*Number of organisers*
Santiago			
1. Santiago	*Fiesta*, procession, horse race, ball	25 July	4
Concepción			
2. The Immaculate Virgin	*Fiesta*	8 December	6
Santa Ana			
3. Santa Ana	*Fiesta*	26 July	6*
4. The Virgin of Dolores	*Fiesta*, procession	4th Friday in Lent	6*
5. Lord of Resurrection	*Fiesta*	Easter Sunday	6
6. Lord of Chalma	*Fiesta*, pilgrimage	Ascension Day	21
7. San Pedro	Mass	29 June	2
Altica			
8. The Holy Cross	*Fiesta*	3 May	7
9. San Jerónimo	Mass	30 September	2
10. San Salvador	Mass	6 August	2
Magdalena			
11. Santa María Magdalena	*Fiesta*	22 July	4
Rosario			
12. The Virgin of Rosario	*Fiesta*	7 October	3
San Martín			
13. San Martín	*Fiesta*	11 November	3
14. Virgin of Transito†	*Fiesta*	13 August	6
Tlaxcalchica			
15. The Holy Cross	*Fiesta*	3 May	3
La Villa			
16. Virgin of Transito†	*Fiesta*, fair	4th Friday in Lent	6
17. Virgin of Guadalupe	Mass	12 December	2
Exaltación			
18. Lord of Exaltación	Two *fiestas*, fair	1st Friday in Lent 14 September	21
19. Holy Cross	*Fiesta*, long procession	3 May	4
Main church			
20. San Juan	*Fiesta*, bull ring	24 June	5
21. Virgin of Guadalupe	*Fiesta*, bull ring	12 December	4
22. Santo Niño	Several *fiestas*, *posadas*	Christmas time 6 January 2 February	2
23. Santo Entierro	Procession	Good Friday	6*

* The same six men are *mayordomos* of Santa Ana, Virgin of Dolores and Santo Entierro

† The same image

the main church—Virgin of Guadalupe, Santo Niño and San Juan—have an organisation cutting across *campans*. These all-town *fiestas* help to allay the antagonism which exists between people of different *campans* and which is often expressed in verbal abuse. For instance, the people of Santiago-Texcalpa are referred to as *jarreros envidiosos* ('envious jug makers'), alluding both to their poverty—which makes them envious of the others' lot—and to their pottery making. The people of Santa Ana are called *muñequeros* ('doll makers'), because traditional healers—who use clay dolls in certain ceremonies (Rojas, 1973: 246 ff.)—lived there. Those who live in Rosario have a reputation for being rich and mean. However, most people stress that these categorisations are meant to be a joke and do not imply any serious conflict between *campans*,[11] although they often mention the existence of hostilities in the recent past (some thirty or forty years ago). Allegedly, if a young man set foot in a *campan* other than his own—especially by night—he was insulted and even physically attacked by the local youths. The reason was that the visitor might be courting a girl of the neighbourhood. There are no explicit rules for preferential marriage in Tlayacapan (nor in other settlements in the area): it does not make too much sense to dispute over women if land is only inherited by men. But, in fact, there is a certain bias towards endogamy within a *campan*. (For demographic reasons, endogamy cannot be the norm; see Tables 7.1 and 7.2.)

One could hypothesise that marriages between people from different *campans* tended to undermine the closely-knit web of affinal links among neighbours, and that people did not want to lose such a stable, manageable, multifunctional social milieu. This was not a conscious motive. When people threw stones at intruders, they expressed their antagonism in terms of loyalty to saints and carnival parties, or by verbally denigrating members of other *campans*. Such antagonism is mitigated by the all-town *mayordomías*, in which people are given the opportunity of defining hierarchical and co-operative relationships with people from all *campans* and *barrios*.[12]

Fiestas and the agricultural cycle

I have so far discussed *fiestas* in terms of their territorial implications. However, they also mark salient dates in the agricultural calendar. A description of the year's *fiestas* in relation to the agricultural calendar is essential to understand their meaning for people: *fiestas* are not isolated events but their celebration and cycles are interwoven with the flux of everyday life. This perspective also allows us to explain variations in *fiestas* which do not arise from territorial considerations, and to see the relation between *fiestas* and celebrations of other types, such as the carnival. I shall therefore discuss each *fiesta* in relation to its place in the year's agrarian cycle, and interpret the symbolic significance it has for people.

The Holy Cross (3 May). The 'Cross of May' comes before the start of the rainy season. It is the time of the year when the star consellation known as 'the Southern Cross' is visible in the heavens. A shower of rain usually falls about this time, after which there should be no more heavy rain until the middle of June. In the interval, fields are dug over and made ready. It is the time in which ties of friendship and brotherhood—which are necessary for agricultural work—should be 'prepared as carefully as a field, sharpened as a machete' (*los necesitamos barbechados como el campo, filosos como machetes*).

There are three different *mayordomías* of the Holy Cross: the first in Altica (north), the second in Exaltación (south), the third in Tlaxcalchica (east). (There were more crosses in the past, so people say.) The cross of Exaltación attracts people from more than one *barrio*. The *novena*, the mass and the afternoon feast are attended only by *barrio* members; but after the meal there is a procession to plant a cross on the top of the southern Cerro de la Ventanilla (a hill which is visible from all parts of town), and this draws people from other *barrios*. A wooden cross, four metres long, is taken up the hill by a volunteer youth, who may or may not be a *barrio* member; usually, several young men take turns carrying the heavy burden. He is followed by men, women and children, with torches, candles and flowers. The party comes back to town after sunset; their lighted candles can be seen as they descend the hill. After the procession informal gatherings and drinking groups usually continue into the small hours.

The Lord of Chalma (Sunday before Ascension Thursday). This is a very expensive *mayordomía*. The first *mayordomo* is helped by twenty secondary ones, plus 'companions'—who include people from other *barrios*, and even from other *campans*. The *mayordomía* is associated with a nation-wide pilgrimage to the shrine of Chalma, a town in the state of Mexico. The local image is a replica of a statue of Jesus Christ which is kept at the shrine. Tlayacapan is on the pilgrimage route from the south-east—i.e. from eastern Morelos and also from the states of Puebla, Guerrero and Oaxaca. The *fiesta* feast is lavish and continues on the following day (Monday) when pilgrims from other areas arrive. The first *mayordomo* must provide bed and board for everyone—some fifty to 100 or even more. Usually, the *mayordomo* has to provide extra food for the pilgrims, as the Sunday meal rarely proves to be sufficient. The Monday night is spent singing hymns and drinking punch. Old people complain that they cannot go to Chalma any longer and that they are near death; they also tell stories about the Lord of Chalma's miracles. On the Tuesday, at five o'clock in the morning, the pilgrimage continues, augmented by a Tlayacapan contingent (which in 1971 was composed of thirty-seven men). The *mayordomo* walks at the front carrying a banner. He is responsible for the physical safety of his fellow villagers while the journey lasts. They arrive at Chalma on the Wednesday afternoon; but do not go into the shrine until the Thursday morning. They pray and have a mass said for good luck in the sowing and for their families' good health. On the way back, they spend the Friday night at the house of a *mayordomo* in Ixcatepec (near Tepoztlán). They arrive at Tlayacapan on the Saturday afternoon. A welcoming party of neighbours and kinsmen waits for them outside the town. Each Tlayacapan pilgrim is given a garland of white flowers (*cacalosuchil*); in exchange, he gives a picture of the Lord of Chalma to each of the friends and relatives who came to welcome him. It is considered a great humiliation not to be greeted by anyone on return. Everybody goes to the Santa Ana chapel to thank the Lord of Chalma for their safe return. There follows a final meal prepared by the wives (a light meal: bread and coffee). Pilgrims from other places stay the night and continue their journey on the Sunday.

Attendance at this *fiesta* (Sunday mass and meal) is restricted to *barrio* members and to those people from other *barrios* who have given money; but the pilgrimage draws people from all *barrios*. One of the *mayordomos* explained that the Santa Ana neighbourhood was especially beloved by the Lord of Chalma, but the whole town should gain His favour lest the crops fail. Hospitality and inter-village co-operation is seen as a way of earning generous favours from heaven.

San Juan (24 June). This *fiesta* marks the beginning of the sowing. The rains have now started; the ground should be soft and moist—not soggy. If soil is in a bad state, the San Juan statue is taken out of the main church and into the fields so he can see the situation. The San Juan *mayordomo* relies on the co-operation of the whole town to finance a bull ring—which is attended by visitors from other Highlands and even lowland settlements. (Recently, the municipal council has played an important part in organising a bull ring in an attempt to raise money for public works. However, it still relies on the ability of the *mayordomo* to raise funds.) If the town has suffered losses in the previous harvest, there is no bull ring.

San Pedro (29 June). This saint does not have proper *fiesta*, but a mass, attended by a small group of neighbours. The mass explicitly begs divine favour for a good rainy season.

Santa María Magdalena (22 July), Santiago (25 July) and Santa Ana (26 July). These are the *fiestas* of the heavy rains, to ask the saints' protection against dangerous wind and hailstorms. (Between 1 and 20 July the rains are moderate; sowing must be completed by the 20th.) Santiago's (St James the Apostle) *fiesta* is the brightest of the three. The apostle is represented by a life-size wooden statue of a warrior riding a white horse. In his honour, a contest known as 'the horse race' takes place.[13] An evening ball for the whole town is held in the central square. This is a recent innovation, the idea of the municipal council which undertakes the organisation and raises money by selling the tickets.[14] Older *barrio* members think that it is not very respectful to the apostle to have a ball in his name; young people think it should be organised by the *mayordomía*, so that the money could be used to improve the *fiesta*. However, the ball itself has proved very successful: it is an opportunity to relax in the middle of a period of hard work in the fields.

Divino Salvador (6 August). This is celebrated with a single Mass, attended by a handful of devoted neighbours.

Virgin of Tránsito (14 August). The image is a recumbent statue which represents the Virgin Mary in death.[15] She is waiting to be awakened—as the plants are awakened from the ground and rise to heaven. At this time, the rain is easing off—people call this period *la calma de agosto*, August quietness—and plants are sprouting. (Too much rain is dangerous, as the shoots can be scorched by chemical fertiliser.) On 13 August, the *mayordomos*, their companions, wives and relatives spend the night in the chapel of Saint Martin—where the statue is kept—praying, telling stories of the Virgin, and eating: a meal is served by the wives. (The *fiesta*, with its meal and mass, is celebrated on the following day—the vigil of Our Lady's Assumption.)

Lord of Exaltación (14 September). Reputedly, this image of Christ is 'very miraculous'—even more so than San Juan, who is the town's patron saint. It has twenty-one *mayordomos*, who share the expenses of two *fiestas*—one in September, the other in Lent. The September *fiesta* marks the last period of heavy rain. The chapel is full of maize blossoms, which are then painting the fields white and yellow. An all-night vigil precedes the *fiesta*: prayer and penitence are needed to propitiate for a good harvest.

San Miguel (29 September). The saint does not have a *fiesta* or even a mass; but his day is very important in the calendar. In Catholic myths, San Miguel is the angel who defeated the Devil and keeps him in chains. But on his own name day (Michaelmas), Miguel gets drunk—as does everyone else. The Devil gets loose, and may ruin fields and damage houses. To stop him, a cross of yellow flowers (*pericones*) should be put in every plot and every house. The appearance of these bright symbols in the landscape warn the farmers that harvest time is near; special care should be taken of the fields in these last days of cultivation; and preparations made for the harvest.

San Jerónimo (30 September). A mass is said, and a rosary for a good harvest.

Virgin of Rosario (7 October). This *fiesta* marks the beginning of the harvest. To say the rosary brings full maize grains and prevents disease in tomatoes.

In fact, only part of the crop is actually harvested at this date. It is the time of *elotes* (tender maize): farmers invite their friends, *compadres* and relatives to have an *elote* lunch in the field. (The cobs are roasted on a bonfire and eaten with salt, lemon and chili powder.) Sometime during October, a midnight mass (*la misa de las espigas*) is celebrated in a maize field, lit by candles and lanterns. It is organised by the Rosario *mayordomo* but attended by the whole town.

San Martín (11 November). Another harvest *fiesta*. The myth holds that Martin was a soldier who gave half his mantle to a beggar: his statue represents him performing this act. The moral of the story is that one should be generous to neighbours so that the harvest reaches a happy ending.

Two celebrations mark the end of the harvest:

Virgin of Concepción (8 December). Attended only by a limited number of neighbours.

Virgin of Guadalupe (12 December). Most houses have a picture of Guadalupe; it is the most popular name for women (occasionally, also for men). Therefore, many households celebrate name-day parties at this date. A *barrio fiesta* is celebrated in the chapel called La Villa (after the village of Guadalupe, where the Virgin is said to have appeared to an Indian in the sixteenth century.) A town *fiesta* is celebrated for the image lodged in the main church. This latter includes a mass which children attend dressed 'like Indians'.[16] (I tried to find out whether people interpreted the Indian-dress as a symbol of submission to the European religion; but rather they saw it as an evident sign of the fact that they were *not* Indians: they could be such only in disguise.) There is also a bull ring, which is a particularly happy event when the harvest has turned out to be a good one.

Christmas posadas. The prosperity brought about by the harvest is reflected in the Christmas parties. Every family has its private Christmas dinner—as lavish as possible—which is attended by family members currently living elsewhere. The unity of the patriarchal household is restored: married men whose father is still alive are expected to have dinner with him. In each neighbourhood, Christmas is celebrated by the *posadas*. These take place each day from 15–23 December in a different house of the neighbourhood. In the early evening, images of Joseph and Mary ('the pilgrims') are taken in procession through the street; the rosary is said, and carols are sung. Then, the party sings a song asking *posada* (a place to stay the night) for Joseph, Mary and Jesus (who is about to be born). When the singing is over, the party enters the house where *posada* is being given, and receives fruit, candy, peanuts, and a hot drink.

El Santo Niño (several days). At the town level, the Christmas celebrations centre on the image of the child Jesus (the *Santo Niño*) which is kept alternately in the main church, in the house of its *mayordomo*, and in the house of its *padrino*. (Why do other images not have a *padrino* (godfather)? 'Because they've grown up,' I was told.) These two ritual offices are the most prestigious and expensive: it is estimated that each of them costs the equivalent to cultivating one hectare of red tomatoes. There are no secondary *mayordomos*, or companions, to help pay the expenses. And about 1,000 people partake of the Santo Niño meals. These include a light dinner (ice cream, biscuits, soft drinks) at the *padrino*'s house on his nomination day (usually Corpus Christi Thursday); a breakfast at the *mayordomo's* house on 16 December—when the image is taken to the *mayordomo's* house from the main church; another breakfast, again at the *mayordomo's* house, on 22 December (when the image goes back to the church); a full dinner on 6 January paid for by the *padrino* (the image is taken to his house); yet another breakfast charged to the *mayordomo* on 2 February. Other expenses are rockets, music, *escamadas*, and new sets of clothes for the statue of the child Jesus. In spite of the expense—or perhaps rather because of it—the waiting list of Santo Niño *mayordomos* and *padrinos* is filled until 1995. Some people put the names of their children on the list, so that the family gets more chances of participation.

The *mayordomo* and the *padrino* are coached in their obligations towards the Santo Niño by a group of old women called the *tenanches* ('mothers of someone', in Nahuatl). There were eleven *tenanches* in 1971, but I was told that membership was open to anyone with enough devotion to dedicate her life to the image. Out of the eleven, eight were 'ritual specialists' in different *barrio fiestas*. These women sing and dance when the Niño is moved; they clean the church or the house where it is kept, make clothes for it, and help the *mayordomo*'s and the *padrino*'s wives to serve meals without charging anything for their services.

The carnival. The carnival is celebrated in February just before Lent begins. 'For us, Lent is certainly a time to do penance. The dry season gets dryer and dryer, hotter and hotter. The *jagüey* water gets scarcer and dirtier. And we are running out of money from the harvest.' The carnival festivities mark the transition between the rainy season—when working life is centred on the Highlands—and the dry season—when work is to be found in the valley down below. The family has saved some money to buy new clothes for everyone and to have some enjoyment—food, drink and a dance—at this time. The biggest market day in the whole year is held on the Saturday before carnival.

The main organisers of carnival festivities are called *autores* (authors) because one of their duties is to write poems about the carnival, about amusing incidents in Tlayacapan, about girls and—very circumspectly—sex. Sometimes these poems may refer to a serious matter—for instance, in 1971 water scarcity and the recent drilling of a well were mentioned. The *autores* take office on 31 January, when three groups of men (one for each carnival 'authorship': Santiago, Santa Ana, Rosario-Exaltación), disguised in masks, funny hats, and old clothes, wander about the streets of their *campans*. The new *autores* carry enormous red banners emblazoned with the name of the *comparsa* (carnival party) in white. (The names of the *comparsas* are: Union for Santiago, America for Santa Ana, Azteca for Rosario-Exaltación.) On the Friday before Ash Wednesday, a bull ring is organised on the central square. Bull rings also take place on the Saturday and Sunday. On the last day, people stay in the square until the evening, long after the ring is over. At about seven p.m., the three *comparsas* come to the square: the *autores* are followed by a brass band and a group of *chinelos* (carnival dancers). The bands play the carnival tunes—which may well be 400 years old; the *chinelos* start the *brinco* (literally: jump), and the whole town joins in[17]. The 'jump' may last for about three hours. At ten, the *comparsas* gather in the centre of the square and a more formal ball begins, to the music of an orchestra from Cuernavaca or Mexico City. After the ball—at about three o'clock on the Monday morning—the *comparsas* go back to business: the 'jump' begins again. Each *comparsa* moves into its own *campan* and chooses a house—usually that of a prosperous farmer or shopkeeper, who has been previously warned by the *autores*—to have breakfast in. Dancing and drinking continue inside the house. At noon, a huge meal is laid out at the house of the first *autor*. After this meal, the *comparsas* go to the square and the 'jump' starts again. The second carnival ball begins at ten. Tuesday follows the same pattern: people drink, eat and dance with superhuman energy. The ball on the Tuesday night is outstanding: two, even three orchestras are hired from Mexico City. During the ball, the names of the following year's authors are publicised. At seven, a breakfast is given by the three first *autores*, after a final 'jump' through the main streets. A sudden chime of bells from all the chapels marks the beginning of Lent. Exhausted and sleepy, people go to the main church to have their foreheads smeared with ash (the Catholic custom), and then go to bed.

The role of the *chinelos*—anyone can become a *chinelo* simply by wearing appropriate clothes—is worth analysing briefly. Their dress—colourful tunics, hats, and pink, bearded masks—is a caricature of the European face. Tradition holds that the Indians and *mestizos* donned *chinelo* disguises in order to ape the free behaviour and manners of the Spaniards. This 'ritual of rebellion' reinforced the subordinate status of Indians: it reminded them that they had to look 'white and bearded' in order to behave as they pleased. Nowadays, *chinelo* masks have no ethnic connotations, but they still symbolise freedom and equality. *Chinelos* may dance with the wife of the rich moneylender and joke with his daughter without being recognised. They feign high-pitched voices and make fun of everybody. Participation in carnival festivities is an earned right: *autores* do not have voluntary 'companions', but each citizen is asked to give a fixed amount to pay for the high expenses of the balls. (In 1971, the fee was 200 *pesos* for each man over eighteen.) This reinforces the *chinelos*' feelings of equality and impunity. However, the fact remains that freedom and equality are still associated with exceptional feasting—and with wearing a mask.[18]

The Lent Fairs. The big carnival market is the first of a series of commercial events which take place during the dry season. This period is seen as the appropriate time for fairs and markets to sell the grain kept from the harvest and other local products such as pottery. The two commercial fairs which are held in Tlayacapan are part of a wider network which includes lowland towns, and also *pueblos* in the state of Mexico—Ozumba, Tepalcingo, Amecameca (Bonfil, 1971). Most fairs occur on successive Fridays throughout Lent.

In Tlayacapan, the first Friday in Lent is devoted to the Lord of Exaltación. A pilgrimage (about 100 people) from Texcoco and Chalco arrives early in the morning. The *mayordomos* and *barrio* members welcome them, and a mass, paid for by the pilgrims, is celebrated. The open market is then set up outside the chapel. Many pilgrims, who are also vendors, stay until Saturday—they are allowed to sleep in the chapel—and attend the central square market.

On the fourth Friday in Lent a fair is celebrated outside La Villa chapel; the image of the Virgin of Transito is brought to the chapel for the occasion. There is a midday mass and an evening rosary. The event is organised by people from the town of Tepoztlán, where there is a *mayordomía* of the image. Legend holds that the Virgin of Transito was originally kept in a Tepoztlán chapel, but was taken to a Tlayacapan artisan to be re-painted. When the Tepoztlán *mayordomos* came for it, the Virgin 'would not go': it was so heavy that they could not carry it further than La Villa. So it was left there—and later moved to San

Martín's chapel for safety. The Tepoztlán *mayordomos*, however, could not abandon their obligation and continue sponsoring a *fiesta*—which is also a fair. Another tradition says that the fair marked the ending of the *zafra* (cane-processing period), when the *hacienda* provided gifts (sugar lumps, rum) for its workers. In fact, in the present-day celebrations, the Tlayacapaners perform the Danza de los Vaqueros (Dance of Cowboys), which represents the search for lost cattle by *hacienda* workers.

The Easter ceremonies. The year's cycle is closed by three ceremonies at the end of Lent: the Virgin of Dolores (sixth Friday in Lent), the procession of the Santo Entierro (Holy Burial), and the Lord of Resurrección (Easter Sunday). For the Holy Burial procession a recumbent image of Christ is taken out of the main church into the chapel of Santa Ana, where it is mourned by the *barrio* women. This procession used to be part of a more elaborate ritual which included a representation of Christ's crucifixion, for which people of the Santa Ana *campan*, dressed as Roman soldiers, apostles and the 'Jewish crowd', performed a short play.

Easter Sunday, which is considered to be the main celebration of the year by the Catholic Church, is a rather minor *fiesta* in Tlayacapan: the dry season has not yet finished; one has to wait for the Cross of May for the agricultural cycle to begin again.

Cults of the dead. I have left these cults to the end of this brief description of the ritual year, because they are not explicitly associated with agricultural activities. In a way, the dead (*mayores*) are permanently celebrated on the household altars—which display photographs or souvenirs of family members who have left this world, side by side with images of saints. A daily duty of household women is to make sure that at least one candle is always burning before the altar. But the dead are especially celebrated at the beginning of November. On the night of 31 October the altar is profusely decorated with yellow flowers (*zempasuchil*), a type of bread (*pan de muerto*) which is only made at this time of the year, and children's clothes. Candles are lit at midnight, together with *sahumerios* (incense offerings). Both the candlesticks and the incense holders are made of black clay—a speciality of Tlayacapan potters. On this night, the *muertos chiquitos* (dead children) of the family come to enjoy the offerings. On the following night (All Souls) it is the turn of the adults, for whom new offerings are set out—adults' clothes, more food, candles and *sahumerios*.

In addition to this household celebration, the whole town gathers for ceremonies in the cemetery—which is by the chapel of Exaltación. On the night of 1 November, a brass band stands in the middle of the graveyard, and plays from midnight until two or three in the morning. They do this without payment; it is their service to the town. The sad, beautiful tunes are only played on this day, or at burials—including the Holy Burial procession. While the music is playing, people go to say a brief prayer by the graves of relatives. On the following day, at about five o'clock in the afternoon, all families go to keep company with their dead ones. Tombstones are decorated with flowers, candles and ribbons. At sunset, the cemetery is a sea of candlelight. By nine, the dead are again left on their own.

Many local people see the cults of the dead not only as a sign of loyalty to the family, but also as a sort of commitment to local identity. '*Los mayores nos dieron neustras constumbres*' (The dead ones gave us our customs, our way of life). Ancestors live not only in the graveyard and on the domestic altars, but also in traditional ritual and agrarian wisdom. The unity of the town is not a passing feeling, which finishes after a *fiesta* or a bull ring: it persists over time; its depth transcends the agricultural cycle and reaches far into the distant past—the past of the tombstone, of the dead of many generations joined together in a single graveyard.

Ceremonial activities in other Highland settlements

Before continuing the analysis of ritual celebrations, it is convenient briefly to compare ceremonies in Tlayacapan with those in the rest of the region.

There are many similarities between Tlayacapan and the towns of Tlalnepantla and Atlatlahucan. Neighbourhood festivals alternate with town *fiestas*. There are no clear-cut *campans*, but *barrios* tend to be more precisely defined than in Tlayacapan. Carnivals are organised by several *barrios*. *Fiestas*, though dedicated to other saints (Santiago is a saint who turns up in several *pueblos*), are also associated with agricultural cycles. There are no Lent fairs, but the main end-of-May *fiesta* in Tlalnepantla (the *Cristo Ensangrentado*—blood-stained Christ—celebration) has important commercial events associated with it.

There is a fair in Totolapan on the fifth Friday in Lent, in honour of

Señor Aparecido. But Totolapan does not have *barrio* chapels, or *barrio* saints—although there are May crosses associated with groups of neighbours. (There is in fact no town or village without at least one May cross.) The *fiestas* of Totolapan are for the saints in the main (and only) church, and the priest plays a leading role in their organisation. This means that the emphasis is put more on church events (masses, sermons, communion) than on socialising and feasting. The most important *mayordomía* is that of Guadalupe; the incumbent is always one of the local rich, who obeys the priest's orders.

The smaller settlements each have a chapel, and only one or two *mayordomías*. In addition to the patron saint celebration, there is always a *fiesta* for one of Our Lady's appelations: Candelaria, Concepción, Asunción. For fairs and bull rings, people go to the towns. Larger villages—like San José de los Laureles—have neighbourhood *mayordomías*.

Like the May Cross, certain festivals are celebrated throughout the region: the cults of the dead, the *posadas*, the Virgin of Guadalupe. In Tlalnepantla and San Agustín, there are two additional ritual offices: *fiscal* and *topil*, whose function is to take care of the chapel. These are rotational offices: all men should eventually perform these duties as a service to the *barrio* or *pueblo*.

Variations in the meaning and consequences of ritual

The purpose of the descriptions given above is not merely to display ethnographic curiosities, but to provide material pertinent to testing the hypothesis advanced at the beginning of the chapter, namely, that ceremonial events are open to different interpretations by actors for whom the consequences of such events tend to be different. From the data provided, it is possible further to classify different events according to the inclusiveness of the 'frame of reference' in which actors find themselves. In this context, 'frame of reference' refers to the definition of boundaries which separate participants from non-participants in a given ceremony. Boundary definition can be established at five levels: the household, the neighbourhood, the *campan*, the town, the wider society (Table 7.4). At all levels, the ceremonies imply a dialectical tension between participation and exclusion, between community and differentiation. Differentiation arises in many ways: in division of labour according to sex (for instance, between the *mayordomos* and their wives), in economic stratification (between *mayordomos* and companions), in territorial divisions (between *barrios* and/or *campans*), in selecting

Table 7.4
Ceremonial frames of reference

Level	*Events*	*Actors*
Household	Domestic altar cults	Family members, *compadres*
	Cults to dead relatives	Family members
	Christmas dinners	Family members
	Birthday parties	Relatives, friends
Neighbourhood	*Posadas*	Family, neighbours
	Barrio fiestas	Neighbours (restricted)
Campan	Carnival	Neighbours (extended)
Town	Town *fiestas*	Tlayacapaners
	Chalma pilgrimage	
	Harvest mass	
	Cults of the dead	
Wider society	Chalma pilgrimage	Mexicans
	Lent fairs	

among friends and kinsmen those who become *compadres* and *mayordomos*.

The frames of reference by no means constitute bases for the emergence of corporate or semi-corporate groups—in fact, their members present cohesiveness only in terms of the celebration of a particular ceremony. Not even the household can be strictly considered an organisational unit for extra-domestic purposes. However, the frames of reference do provide a basis for co-operation between individuals otherwise divided by heterogeneous and opposed interests. For instance, most 'companions' are people who have to hire themselves out as labourers even during the rainy season and who, given the abundance of labourers available, risk being subject to frequent periods of unemployment. By developing ritual links with *mayordomo*-farmers, they secure employment.[19] In turn, *mayordomos* feel that they can ask 'companions' to work on their fields without being paid immediately if farming money becomes scarce, or to work an extra hour for free (*faenear*). In short, in the *fiesta* events the relationships of trust needed for agricultural activities are easily established, even between people who, by virtue of their differential access to resources, may have diverging interests in other fields of activity. (For instance, a 'companion' pointed out to me that the current *mayordomo* was paying him very little money for a rented plot, but he did not mind this as they both were united by their devotion to the saint.) Frequently fellow *mayordomos* help each other to fumigate (when it has to be done very quickly) and harvest. However, the performance of a ritual does not suppress the antagonisms which exist in the community: resources are still unevenly distributed; labourer-'companions' compete among themselves for employment; farmer-*mayordomos* compete for prestige. On the other hand, the repetition of the ritual cycle—the nomination of new *mayordomos* both at the *barrio* and at the town level—produces a periodic realignment of heterogeneous interests. The social life of the town can thus be conceived as an ongoing process whereby ephemeral groupings appear and disappear at different levels of inclusiveness.

Antagonism between *campans* does not cease to exist with the performance of town *fiestas*, especially when it involves competition for resources. When the Tlayacapan farming association was founded, the presidency was occupied by a Rosario farmer. Automatically, Santa Ana farmers stopped attending assemblies and participating in association activities: they did not 'trust' the Rosario people. It is also interesting to look at cases in which *barrio mayordomías* have gone to residents in another *campan*. I only encountered two cases (both in Tlayacapan). The first one concerned the 1971–72 *mayordomía* of Santiago. Daniel, the first *mayordomo*, invited his cousin Cleto—who lives in Santa Ana—to be a secondary *mayordomo*. Cleto's friendship was a valuable asset for Daniel: in addition to being reasonably prosperous as a farmer, Cleto is the leader of a band and has links with all local musicians; he teaches music in the secondary school (although he does not have any formal teacher training) and some people call him 'teacher'. Cleto defended his nomination before the Santiago neighbours saying that his mother and his wife were both born in Santiago, that he was extremely devoted to the saint, and promised that the music for the Santiago *fiesta* would not cost much. The neighbours accepted his appointment, but the old people were uneasy about it. The second case involved open conflict. Ambrosio, a man from Santa Ana, has for years given money to the *fiesta* of the Lord of Exaltación. He asked for the first *mayordomía* year after year, saying that he had made a vow to the Lord of Exaltación because the Lord had saved the life of both his father and his mother. Finally, Ambrosio was given the *mayordomía* by an incumbent who was his *compadre*. There followed several heated arguments with the neighbours, many of whom saw this appointment as totally wrong. ('If the *mayordomía* goes to Santa Ana, our *barrio* will be damned.') Things were further complicated by accusations made by

Ambrosio that previous *mayordomos* (excluding his *compadre*) had mishandled *barrio* funds. As it happens, Exaltación is the richest chapel, because of the alms and contributions left by the Lent pilgrims—which are administered by the *mayordomía*. Ambrosio claimed that the money had been used by some *mayordomos* for personal ends. Eventually a violent encounter took place, in which Ambrosio defended himself from physical attack with a knife—and wounded two men. He left town; there is no longer any question of his being first *mayordomo* (another man took his place). The moral of the story (for the Exaltación neighbours) is that the Lord punished the *barrio* for putting His *fiesta* in the hands of an alien.

The *fiestas* and most celebrations mean money for some people. Musicians—the Tlayacapan brass bands, the Nepopoalco flute players and drummers—are absolutely necessary and have to be paid. Local shops sell a large part of their stock of beer and spirits for *fiestas*, as well as clothes and personal items: 'A *fiesta* is not fun if one does not wear something new'. Rocket-makers sell huge quantities of rockets: no *fiesta* is satisfactory if rockets are not fired at least every fifteen minutes from dawn to the end of the mass (sometimes from the previous evening to the end of the feast). Ritual specialists are paid; they also provide the *escamadas* (coloured candles), which cost 75 *pesos* each, special sweets for the meals, prepare the punch, produce garlands and festoons, and sell rosaries and prayer leaflets. Thanks to their common association—the *tenanches*—they help one another in providing these services to different *barrios*, and share the payment.

The fact that money has to be disbursed entails not only trust, but also administrative ability on the part of the *mayordomos* and *autores*—particularly the principal office holder. Thus, another consequence of local ceremonies is the opportunity they present for the exercise of leadership. Other activities—agriculture, for instance—do not provide any basis for a leader to emerge from a relatively large number of people. Leadership comes about when there are opportunities for organising resources and co-ordinating behaviour in implementing common goals. Small-scale farming obviously provides very limited opportunities of this type. In contrast, the performance of a ritual such as the *fiesta* focuses the attention of a number of individuals on a common goal—the ritual itself—and attracts their co-operation and support. In addition, as the ceremony proceeds, people are able to evaluate the leader's behaviour: his wisdom in the selection of secondary *mayordomos*, the number of 'companions' he attracts, his honesty and efficiency in administrating the finances and, generally, the splendour or dullness of the *fiesta* itself.

In fact, many incumbents of political offices have behind them a past of involvement in *mayordomías*. This means that a man can become a prestigious ritual leader, and 'convert' (in Barth's sense) this prestige into a political asset. At the same time, the ceremonies have an intrinsic political dimension: they provide specific ways of controlling resources which are important in local terms. The civil authorities use ceremonial days as occasions for raising funds for the municipality—for instance, by organising a *fiesta* ball or running a bull ring. *Fiestas* also have a political meaning *vis-à-vis* the Church: they define the terms of religious activity and—most importantly—the use of resources for religious ends.

In the following chapter we shall examine the ways in which Church and State attempt to impose their own definitions of social hierarchy and alliance, which imply alternative priorities for the use of resources.

Notes

1. The characteristics of 'plural' societies can be summarised as follows: (*a*) lack of common social will among the component sections; (*b*) institutional and/or cultural separation; (*c*) the existence of a dominant elite which holds the power and social authority, and binds the different segments together; (*d*) the existence of a technological gap between the dominant and the other segments, which results in the co-existence of at least two different types of economies (see Furnivall, 1948; Boeke, 1953; Smith, 1965.

2. Gluckman's analysis of colonial Zululand as a single social system becomes relevant here (1958). See also Kapferer's critique of Smith's use of the concept of the plural society (1966).

3. 'The higher the social status of the speaker the more people he regards as "really Indian"' (Pitt-Rivers, 1967: 75).

4. In San José de los Laureles, for instance, people over sixty speak Nahuatl among themselves and to their children. The next generation (forty to sixty) speak Nahuatl to their parents, but Spanish among themselves and to their children. People under forty usually speak Spanish only. The expansion of the national school system and the intensive contacts with external markets have also been important factors in the adoption of Spanish.

5. One ravine clearly separates Santa Ana from El Rosario; another Santa Ana from Exaltación; a third, Exaltación from El Rosario. Santa Ana is separated from Santiago by a paved road.

6. There are many problems with this concept of ritual; but this definition suffices for present purposes. The sanctions I am referring to are put in terms of offences to God, the Virgin Mary or the Saints.

7. Even trivial behaviour like drinking becomes a sort of ritual occasion. People are

expected to drink at exactly the same rhythm as the person who is paying the bill, or who owns the house in which a gathering takes place. The rhythm is usually indicated by toasts. This allows people to exchange abundant information and reach a degree of emotional attachment to each other. 'We're not anonymous, but public drunkards', said a farmer when the Cuernavaca Alcoholics Anonymous agent tried to recruit members in Tlayacapan.

8. It has been said that redistribution of wealth (resulting in social homogeneity) is a primary explanatory variable for *fiestas*. (See above for my critique of the model of the corporate community.) Redistribution of wealth is *one* among many possible consequences of a *fiesta*, but it is not methodologically valid to substitute a consequence for a cause.

9. It is interesting to note that one of these men was an active *mayordomo* in his youth, before becoming a 'big shot'.

10. In some *mayordomías* there is a waiting list of candidates which is kept by the incumbent.

11. One could compare this to 'joking relationships' among social categories whose members stand in relationship to one another both as competitors and co-operating partners, as described by Mitchell (1957).

12. Oral tradition holds that these big *mayordomías*—certainly the San Juan *mayordomía*—were in the hands of the people of the central square. This opposition between 'the centre' and the *barrios* is also found in other Mexican towns—such as Cholula (see Bonfil, 1973: 277 ff.).

13. In this contest competitors on horseback have to put a stick through an iron hoop suspended on a ribbon. The ribbons belong to girls of the *barrio* and the rider who succeeds in spearing the ring and the ribbon has the right to dance with its owner in the evening ball. The *mayordomos* organise the contest. The occasion does not afford brilliant displays of horsemanship: the horses are old and the riders rather awkward (as one would expect in a region with no tradition of cattle-raising or riding); but it attracts a relatively large audience from all *barrios* and even from outside the town.

14. Balls usually take place on the market platform which is fenced off with a rope so that only ticket holders can get in.

15. According to one Catholic tradition, Mary died a natural death, and shortly afterwards she was resurrected and raised to heaven by her son Jesus Christ.

16. The boys are supposed to represent Juan Diego—the Indian to whom the Virgin Mary appeared—and the girls Malinche—Cortés's Indian mistress and interpreter. Compare the description of the Guadalupe *fiesta* in Hueyapan in Friedlander (1975: 104 ff.).

17. The *brinco* is a complicated dancing step whereby rows of six or eight dancers (men and women) move forwards and backwards. The National Anthropological Museum of Mexico has produced a record of the Tlayacapan carnival music, played by the original brass bands.

18. Bands of *chinelos* are also found in Yautepec, Oacalco, Oaxtepec, Casasano, Cuatlixco and Tepoztlán.

19. In some *mayordomías*, the people who usually contribute small amounts are categorised as 'the fraternity' (*hermandad*), and the people who have been *mayordomos*, as 'the council' (*cabildo*).

VIII
ON CHURCHMEN AND POLITICIANS

Los padrecitos nos prometen el cielo. Los políticos también. [Angel Rojas, Tlayacapan farmer.]

The ceremonial activities described in Chapter VII present symbols and beliefs which have a variety of origins. The saints, and many of the legends attached to them (San Martín and the beggar, for instance), belong to a Christian tradition. Some beliefs—for instance, the faith in the magical properties of yellow *pericón* flowers—probably come from the prehispanic world. Other beliefs are less clear in their origins: the ceremonies of the dead, the agrarian interpretation of the calendar. One is tempted to describe this multiplicity in Redfield's terms—i.e. to speak of local culture as a 'way of life composed of a Little Tradition of the non-literate and a Great Tradition of the literate and philosophic few' (1964, vol. I: 120). The problem with this approach is that it assumes that each tradition is somehow internally consistent and homogenous. But they are not. The meanings of folk ritual are variously interpreted according to the actors' social context and attributes. Similarly, the 'Great Tradition' of Catholicism has adopted very different attitudes in regard to village festivities according to the political position of the Church at different periods and to the social situations in which Church representatives are involved.[1] The Church used *fiestas* in the colonial period as a means of enhancing its control over the population; lately it has become openly hostile towards these events which are now to a large extent independent of clerical control. However, there have also been cases of priests who have manipulated traditional festivals to their own personal ends. A fruitful way of looking at religions syncretism is to examine how different actors stress or have stressed different symbols or beliefs in specific circumstances.[2]

The folk interpretation approves the use made by local actors of a culturally inherited set of symbols—including formal sequences of expressive behaviour—for the creation and maintenance of relationships of trust. These symbols are also used by other actors—the priests—but often in order to impose another interpretation,[3] one in which they can achieve and maintain a superordinate position which might then be 'converted' into assets in other spheres of social life.[4] The first part of this chapter is devoted to examining the role and strategies of the Church and the priests. Although priests are in certain instances the most powerful individuals in the region, their actions are heavily constrained by social forces, often resulting in serious conflicts and oppositions. It will be argued that conflict and opposition are not merely unintended consequences of Churchmen's activities, but are rooted in the very nature of local politics, which can be characterised as 'fragmented'. Finally, the fragmentation of local power domains will be related to the centralisation of power, at the national level.

Folk religion and Church politics

In popular horror tales of the Morelos Highlands, the Devil is frequently depicted as a tall man who dresses as a Catholic priest, in black cassock and bonnet. In Atlatlahucan he is said to emerge from an *amate* tree, and attack people and cattle. On the other hand, some stories tell of the Devil's generosity towards his protégés, whom he is supposed to lead to hidden treasures of gold and silver. The image of the priest is thus associated with a powerful mystical force which can be both benevolent and malignant. There are numerous stories about priests who seduce villagers' wives and daughters, steal valuable objects from churches and chapels, and demand exceedingly high fees for their ritual services; but there are also many stories about priests who lead a holy life, give money to the poor, and work miracles.

This ambivalence towards the representatives of the Church is rooted in historical tradition. People refer to the missionaries and to the clergymen who began the independence movement as heroes; but local tradition regards the priests of the period immediately before the revolution as allies of the *mestizo* elite. In fact, at the beginning of the twentieth century the priests had managed to recover many of their privileges by building up personal alliances with the upper classes. Local bosses permitted the collection of tithes,[5] and the Church attempted to regain control over religious ceremonies. Thus, folk religion became an expression of the opposition of the lower strata to both the elite and the priests. To this day many people say that they have kept their religious faith in spite of the efforts of the clergy.

The present-day attitude of the clergy towards the *fiestas* is generally hostile; again, this is related to their political position in the country. The Church did not show any sympathy to the 1910 revolution, or to the subsequent land reform.[6] The post-revolutionary constitution of 1917 was, in many aspects, openly anticlerical.[7] In some parts of Mexico, the Church was openly persecuted by the Government, and, from 1926 to 1929, official Catholic worship was suspended throughout the country. Later, in the 1930s, the government took a more tolerant attitude towards the Church, although the anticlerical laws remained unchanged, and persist until now. From 1940 onwards the Church has tried to regain its influence, and to shake off the reputation of being on the side of the rich. One important strategy has been the creation of associations for laymen combining religious instruction and social-welfare work.[8]

In Morelos, local priests are recruited from the urban population. They contrast markedly with the priests of western Mexico, who come from rural families and maintain deep emotional roots in the countryside. I never met a Morelos Highlander who counted a priest or a nun among his relatives; whereas it is difficult to find the same in Jalisco or Michoacán.

Changes in Church ideologies. In 1963, Fr Emile Pin, a French Jesuit who held the chair of sociology of religion at the Gregorian University of Rome, published a report on Catholicism in Latin America. He stressed the superstitious character of 'folk religion', and pointed out that *fiestas* encouraged gambling, dancing and heavy drinking rather than prayer (Pin, 1963: 62–3). He argued that the mental confusion of the lower classes, particularly 'the peasants', leads them to identify the images of the saints with pagan deities. The relationship established between the saint and the worshipper is conceived by the latter as a magic contract which obliges the saint to bestow favours on him—usually 'material' favours, such as health and money. One essential task of the Church should be to change this 'magical' view in order to create a truly Christian ideology in the people—one in which the believer sees himself, by virtue of his faith in Christ, committed to the transformation of the world in order to achieve a just society, based on love and peace (Pin,

1963: 56–7).

Pin's book came out in a period when the clergy and the Catholic urban elites in Mexico were giving a great deal of thought and discussion to the role of the Catholic religion in society, stimulated by the wind of change brought about by the Second Vatican Council and the open attitude of Pope John. In spite of the resistance of some clergymen, policies were devised to bring changes in the liturgy, and to reorganise religious instruction. There was much talk about 'going back to the sources'—i.e., reading the Bible and paying less attention to pious books and catechisms—about giving less importance to external forms and more to inner beliefs, and about the Christian concern for social justice. The role of the Church—it was stressed—should be one of service and not of dominance. The attitude to other religions—especially to the Protestant sects—should be one of fraternity and not of competition.[9]

The pioneer and leader of the movement for Church reform in Mexico has been Sergio Méndez Arceo, the bishop of Cuernavaca—the diocese to which the parishes of the Morelos Highlands belong. He was the first to introduce important changes in Church ceremonies, including the use of Spanish instead of Latin. He refurnished his cathedral and removed the images of the saints so that worship would be centred on Christ. He has repeatedly spoken against those religious practices which he classifies as 'superstitious and alienating'.[10] He is also known for his sympathy towards socialism and his critical attitude to capitalism in Mexico. Recently, he met with some hostility from the government over his public defence of workers' strikes in Cuernavaca[11] (Rascón and Cordera, 1972).

The role of local priests. Even when it defends social causes, 'progressive' Catholicism remains basically individualistic and authoritarian. The concern for social justice emerges from individual persuasion and from doctrinal definitions given by the bishops, the parish priests and the Vatican Council—not from the consensus of a local community. In contrast, the beliefs manifested in *fiestas* could not persist without the support of the group as a whole. At the local level, the possibilities for manipulation that such individualistic, vertical ideologies offer a priest are many: the faithful should depend on him for the correct development of their personal consciences. However, in the Highlands, priests found that having a 'correct' conscience was irrelevant for most people. The elders, and not the priests, are consulted about beliefs and traditions. The priest is regarded as an *ad hoc* functionary for the performance of baptisms and marriages—which are occasions for *compadrazgo*—funerals, and for *fiesta* masses. People rarely go to confession and communion. Sunday masses are attended mainly by women for whom there is a tradition of prayer which does not exist for men. (It should be remembered that it is women who take care of *barrio* chapels and organise *barrio* rosaries for the saints.) If a *mayordomo* 'lives in sin' with a woman, nobody sees this as inconsistent with his religious office. (In fact, many *mayordomos* are in this situation.) In short, the Highlanders are highly 'heterodox and immoral'.[12]

For most clergymen, this indifference to Church authority stems from the sheer ignorance of the peasantry. The predictable reaction is often to attack the practices which the clergy perceive as sustaining this ignorance and which, at the same time, have the effect of relegating their own role as priests to a secondary position in the local ceremonies and in local social life generally. A strategy used by the priests has been to found associations which they hope will foster 'proper' religious instruction and become a substitute for *mayordomías*.

Fr Pablo, the Tlayacapan parish priest from 1960 to 1968, founded the Adoración Nocturna (Night Adoration) for men, and the Vela Perpetua (Constant Vigil) for women. The first was very successful especially when Pablo began to extend its scope to non-religious activities: he was able to give free medicines to its members thanks to his connections with Mexico City laboratories, and organised a co-operative to buy fertiliser. At the same time, he put pressure on the members and demanded that people should attend mass and generally practice 'orthodox' religion. In sermons, he accused *mayordomos* of using sacred places—the chapels, the chapel yards—for profane activities such as drinking, eating and playing music. He refused to say mass in a *fiesta* unless the *mayordomos* agreed to suppress certain events—for instance the firing of rockets, music, and bull rings. People were shocked at this. As one *mayordomo* put it, 'If the priest is paid for the mass, why should musicians and rocket-makers starve?' However, several *mayordomos* who were members of the Adoración Nocturna complied with Pablo's requests. This caused quarrels and discontent. *Fiestas* were cancelled because of disagreements among the organisers. Pablo also tried to prevent organisers from tolling the bells on the days of big *fiestas* by releasing an enormous dog on the roof of the main church. For many people, the priest was truly antireligious. A committee complained to bishop Méndez Arceo, who replied that Pablo was following the line of the Vatican Council; and that the high expenditure on spirits for *fiestas* only benefited the exploiters of the masses. Some Tlayacapaners thought that Pablo was really after the money spent on *fiestas*. He was always complaining—in his Sunday sermons—of the paucity of the alms received by the Church. Finally, several parishioners publicly accused Pablo of seducing a woman. The bishop declared that he believed in the priest's innocence; but, to avoid scandal, Pablo was sent elsewhere.

Pablo's case shows how a series of conflicts directly related to religious matters can develop. However, in other cases the conflicts have transcended the religious sphere and unchained disputes in other domains. I consider it convenient to look at two of these in some detail.

Religious conflict and political oppositions

Religious–political dispute in Atlatlahucan (1966–70)

The story begins with the death of the parson (*párroco*) of Atlatlahucan in 1966. As there was a shortage of priests in his diocese, the bishop of Cuernavaca asked Father Wilson, at the time parson of Tetelcingo, a lowland community near Cuautla, to take care of Atlatlahucan in addition to his own parish. Wilson was an American priest who had come in 1960 to work as a volunteer in rural Mexico. Tetelcingo is a rather poor village; but Wilson received additional income from the United States.

The bishop later decided that Wilson should live in Atlatlahucan—the larger of the two parishes—and visit Tetelcingo once or twice a week, but when Wilson was about to move, the parishioners of Tetelcingo locked him up in his house for several days, until the bishop reversed his order.[13] Wilson agreed to stay in Tetelcingo and visit Atlatlahucan as often as possible.

The people of Atlatlahucan were disappointed at this decision. They resented the slight to their town. They had prepared a big welcome party, which was wasted. There were already rumours about the strange ideas of the bishop and the younger priests, and it was common knowledge that the saints had been removed from the cathedral of Cuernavaca. However, Wilson continued visiting Atlatlahucan regularly without meeting any problem, though he did not care much for saints or *fiestas*.

The first conflict between the American priest and some of the parishioners occurred early in 1968. At the time, as there was no resident priest in the town, the main church and the monastery were taken care of by an *ad hoc* local committee, whose members sent occasional reports on the state of the building to the Ministry of the National Patrimony in Mexico.[14] Wilson decided to make alterations without consulting the committee; he built partitions in some of the monastery rooms and cut down a clump of banana trees to make room for a small cistern in the yard. The committee disapproved of this, but Wilson paid no attention. He announced that he intended to make the rooms available to the local schools. The committee then wrote a formal complaint to the Ministry, which was never answered, and the priest went on with his works. He removed the images of the saints from the church. He found a gigantic crucifix in the vestry, and thought it would look imposing suspended by a rope from the roof over the vast space of the aisle—as in some modern urban churches. When Wilson suspended the crucifix, the committee accused him of 'hanging Jesus Christ', and several of their members threatened to cut it down. Wilson then resorted to Pancracio, the municipal president, who took sides with the priest and threatened the disturbers with jail. Later, the priest realised that many people were genuinely scandalised, and the image of Christ went back to the vestry.

This incident was then used by a local political figure, a man called Alfredo, who was a personal enemy of Pancracio. Alfredo had—and still has, as he himself openly admits—grand political ambitions; his aims are to be elected deputy for the district, and ultimately to achieve a major office in the state government. His family has been active in local politics for many years: Alfredo's father was one of the bosses who organised land invasions in the 1920s and '30s. From 1950 on, Alfredo has occupied a number of offices in the municipality, from treasurer to councillor and municipal president. When Pancracio took office, Alfredo was the secretary of the *municipio*; later, Pancracio accused him of dishonest dealings and forced his resignation. Thus, when Pancracio supported Wilson in his quarrel with the committee, Alfredo and his followers spread the rumour that the priest and the municipal president were 'Protestants', and that they wanted to impose a new religion in town.

A few weeks passed. In the Lent of 1968, a group of theology students from the bishop's seminary went to Atlatlahucan to help Wilson in a preaching week. They were well received, and one of their main activities was to distribute modern Spanish translations of the New Testament and to encourage its reading and discussion. After the students returned to the seminary, Wilson left the town for Holy Week and a Dominican priest arrived to celebrate the ceremonies of Maundy Thursday and Good Friday. The Dominican was not a subject of the Cuernavaca diocese; he lived in Mexico City and did not particularly like Méndez Arceo's ideas. He happened to see one of the New Testaments, which turned out to be a Protestant edition, and publicly denounced the book, saying that Catholics were only allowed to read Catholic editions of the Bible.

The Dominican went back to Mexico City. By now many people had begun to think that Wilson, Pancracio, the students and perhaps the bishop himself really were Protestants. Several facts gave credibility to this opinion: Wilson was an American; in addition, Protestants are known by their fondness for reading the Bible. Many local people believe that the Bible is a Protestant book. Two of Alfredo's followers went to see the parish priest of Cuautla and told him that they did not want Protestant hangers of holy images. Some people, incited by Alfredo, even burnt their New Testaments.

When Wilson came back, he was extremely angry. He tried to explain that the New Testament edition had been authorised by the bishop; although it was published by a Protestant agency, it was a very good translation, written in plain language and accessible to everybody. Some people were convinced by this argument, but the campaign against Wilson and the municipal president continued. Wilson decided to visit Atlatlahucan no more, and the bishop appointed a new parson.

The new priest, Father Efraín, was a tough man, who did not hide his contempt for folk religion. He gathered around him a group, and organised Bible discussions and modern liturgical celebrations. The latter included masses in the homes of his friends. Many of the new priest's supporters were also friends of Pancracio. Alfredo's group avoided Efraín, labelling him and his followers 'the evangelists'. Alfredo began to mobilise support among the groups involved in the organisation of *fiestas*, who felt rejected by the priest. Alfredo told them that Efraín, Pancracio and their supporters wanted to get rid of all the holy images and to turn the church and chapels into dance halls.

In addition, Alfredo found another way of attacking Pancracio. He accused him of exploiting the sand pit situated on the town's communal land for his personal benefit. Whether this accusation was true or not was never clear; but the Department of Agrarian Affairs intervened, and Pancracio resigned from the municipal presidency.

However, as his term was due to finish in a few months, he had arranged with his friends in the P.R.I. that the next candidate for the presidency should be his nephew, Manuel, who was a friend of Efraín's.

Manuel took office, and his election was confirmed by the Junta Nacional de Municipios. But shortly after, Alfredo and a crowd of his supporters occupied the town hall and declared the election invalid, on the grounds that Manuel did not have the support of the majority. They proclaimed a friend of Alfredo's as the new president, and this was confirmed by the governor of the state. The federal office of the P.R.I. did not agree with this appointment, but did not intervene to remove the new president. As a result, there are two municipal governments in Atlatlahucan, each claiming to be the legal authority, and each having their own supporters. In practical terms, this means that most people register their marriages, the births of their children, and the deaths of the family with one or other party. A number prefer to go to other *municipios* to register, to avoid any problems.

A few weeks after the town had been thus divided into two political segments, Father Efraín left Atlatlahucan and went to live in San Miguel Tlatetelco. He says that he left because of the threats and insults he received from Alfredo's friends. The latter claim that Efraín left because he did not like the Catholicism of the people. Nowadays, a group —'the evangelists'—go to Tlatetelco every Sunday to mass, and Efraín visits them occasionally—though he refuses to set foot in the main church. This has again been taken over by the church committee—Junta de Vecinos—and the images have been restored.

Comments The initial cause of the chain of events just described is probably Wilson's definition of the situation in terms which did not take into account local standpoints. He did not attempt to negotiate or persuade anyone, but remained a rather cold figure throughout; his only outburst was when the New Testaments were burnt by Alfredo's followers. Basically, Wilson was not interested in establishing a personal superordinate role in Atlatlahucan. He already had a secure position in Tetelcingo. He did not need the money from religious services: his U.S.A. allowance enabled him to operate with a high degree of freedom. He was a foreigner: unlike Mexican priests, missionaries do not have a pressing problem of identity in a society which is anticlerical at a national level.

Wilson's attitude contrasts with Pablo's: the latter had to find a way of securing an important role in Tlayacapan. Similarly, Efraín adopted an active position in relating to the Atlatlahucaners. If Wilson's actions had made the local church committee practically redundant simply by ignoring it, Efraín openly showed his dislike of traditional practices. He even tried to change some important rules of the game by moving the place of ritual out of the church and into private houses. To celebrate masses in houses became a rather common practice among 'progressive' priests in the late 1960s. The purpose of this was openly anti-ritualistic: religion should not be a matter of routine, but of deep personal persuasion. Efraín's aggressive strategies attracted support, and also deep antagonism. An intelligent, cultured, vivacious man, the priest established friendship links with some of the better educated farmers and with the teachers. His enemies were the storekeepers and moneylenders (not all of them, though), the people from the local church committee, and many *mayordomos* and 'companions' who felt attacked and rejected.

It is significant that Wilson, Efraín and their friends were accused of being Protestants. In Atlatlahucan, as in most Morelos towns, there is a small group of Protestants. They are usually aloof from local social life; but the rest of the people are aware of their existence and of their religious practices. They are associated with innovation, broken traditions, foreignness, bizarre rites, Bible reading instead of catechism memorising. Some have unusual occupations, such as poultry farming or photography. We saw that Catholic folk *fiestas* provide a convenient social milieu for developing relationships which can be also used for agricultural entrepreneurship. Similarly, conversion to Protestantism not only facilitates co-operative relationships which can lead to opportunities for new occupations, but also provides the rationale for eschewing traditional ritual.[15] Thus, the very word 'Protestantism' evokes hostility and mistrust: Protestants are people who never help in communal matters.

However, cleavages within the town cannot be understood simply in terms of religious disagreements, although religion is used as a social idiom to express antagonism. It is also—and fundamentally—competition for political power which unchains antagonism and strife. If a priest has difficulties in defining his role in a rural area—not only because of the Church's institutional weakness, but also because his services may be largely irrelevant—a local politician is also in an awkward position: institutional political power is mainly vested in the federal government. People who are interested in politics rely on municipal politicking as a springboard to reach higher levels, or else as a means of gaining access to certain resources. For a man like Alfredo to be excluded from an office—after many years of hoping for a higher post—could mean the end of a career. In his turn, Pancracio felt uneasy having Alfredo as municipal secretary. Pancracio is rather young (thirty-eight) and politically inexperienced; Alfredo is older (fifty) and on friendly terms with many politicians in the state and federal government (though not with anyone important). After removing Alfredo, Pancracio supported Father Wilson against the church committee without realising the implications of his action. Later Alfredo's accusations had the effect of uniting Pancracio's followers with Wilson and later with Efraín. Progressive and traditional Catholicism were used as banners for expressing differences between political factions.

What may seem surprising is the lack of intervention on the part of higher authorities to prevent the emergence of two different municipal councils. But the 'divide-and-rule' strategy has been one of the most important characteristics of post-revolutionary federal government action.

Religion and water in Tlayacapan

In 1970, the town of Tlayacapan faced a rather unusual situation: it had three resident priests. (Most *municipios* in Morelos have only one each.) The first was Mamerto, the *párroco*, who from his arrival in 1968 met difficulties similar to those encountered by Pablo. Mamerto had the bright idea of knocking down a set of baroque lateral altars in the main church. The altars were indeed ugly, and did not match the austere fabric of the Augustinian building. But several images stood on them, and Mamerto had to face an angry group of *mayordomos* and devotees, one of whom threatened the priest with physical violence. Mamerto has since remained in the background and sometimes refuses to participate in *fiestas*.

The second priest is Fabrizio. He is a man in his early forties who arrived in Tlayacapan in the winter of 1969. In addition to his ecclesiastical office, he has a degree in architecture and has designed churches and houses in several parts of Mexico. He helped in the modernisation of Cuernavaca cathedral, and is a personal friend of the bishop. He is also a painter and sculptor, and has enjoyed some fame in artistic circles. Through his family and his professional activities, Fabrizio is well connected. Before settling in Tlayacapan, he had lived mostly in Europe for the previous ten years. In 1969, he decided to go back to Mexico for good and to settle in a village where he would have time for his artistic work and at the same time undertake social projects in the locality. As a priest, he had to be dependent on a bishop; and he wanted to be in a diocese where the bishop would not object to whatever activity he wanted to undertake. So he chose the diocese of Cuernavaca. The bishop suggested several villages within the diocese where he could live. Fabrizio chose Tlayacapan, because of its natural beauty, and because it was in easy reach of Mexico City.

In Tlayacapan, his first visits were to Mamerto, and to the *ayuntamiento* (municipal council). The parish priest had already been told by the bishop to treat Fabrizio well, and he agreed to introduce him to all his acquaintances. It should be noted, though, that Fabrizio had not been charged by the bishop with any parish duties. The *ayuntamiento* told Fabrizio that the people needed water, a secondary school, and a clinic.

Fabrizio began to recruit some of his personal friends in Mexico City to form the staff of the school. The school was to be a private one, and the ownership would be vested in the parents, who would have to collect fees from the students to cover expenses. The salaries would be very low, and that was why Fabrizio wanted his friends to volunteer for the job. Once the school was established, it should be easy to persuade the government to take over. Fabrizio's friends arrived: a priest, with experience of school work in rural areas (Humberto); a Spanish lawyer and his wife; a lecturer in mathematics and a girl who soon became his wife. In addition, Fabrizio persuaded a young doctor to come to live in the town and open a clinic. The clinic was also to be private, but the doctor agreed to charge minimal fees. The doctor and his wife also agreed to teach in the school. The whole group was well settled in Tlayacapan by May 1970—six months after Fabrizio's arrival. They did not form a close-knit unit and lived in separate houses, but people called them 'the professionals', or 'Father Fabrizio's family'. In addition, writers and artists from Mexico City and Cuernavaca came to lecture in Fabrizio's school.

Meanwhile, the Tlayacapan people had found out that Fabrizio was much more accessible than Father Mamerto concerning matters of traditional religion. He and his friend Humberto (the other priest) attended *fiestas*, and agreed to perform any ceremony requested. Fabrizio even agreed to take part in a ceremony traditionally performed by healers—the 'fright' ceremony, intended to cure children of insomnia, fever and symptoms attributable to acute anxiety.[16] What Fabrizio did was to adapt a ritual approved by the Catholic Church to bless the children. But in the ceremony, he used a white flower, locally said to have magical powers, to sprinkle holy water on the child's head. Soon Humberto adopted the same method.

People thought Fabrizio was rich—his house was not luxurious, but it was full of apparently expensive items, such as books, drawing tables, carpentry equipment, canvases, sculpting material, etc. In addition, Fabrizio was visited at weekends by some of his Mexico City friends, who parked their cars outside his house. So people started asking him to lend them money; Fabrizio was willing to comply if he had the money at the time, and the amazing thing from the standpoint of the local people was that he did not ask interest. Using his connections, he helped several youths to find industrial jobs in Mexico City and Cuernavaca. In April 1970 Fabrizio and his followers were presented to the town by the *ayuntamiento* at a public meeting. The municipal president stated that the group had come as volunteers to co-operate with the local people. He announced that the clinic was already functioning, and that poor people would have free services there, although a fee would be asked from those who could afford it. Also, a secondary school would be functioning in the town from the next autumn.

By this time, Fabrizio's reputation was consolidated. People thought of him as being well connected in politics. When a dispute broke out between the communal land officer and the municipal president, both men came, separately, to Fabrizio and asked him to settle the case in their favour: Fabrizio said that he had neither the authority nor the power to do it. Later, two police constables asked him to get them government posts as tax inspectors for the state; again Fabrizio had to say that he could not do anything about it.

Fabrizio's most notable feat was the drilling of a well in 1970.[17] Through manipulation of his personal ties, Fabrizio achieved in six months what the municipal authorities could never have done: a loan from a private bank at low interest, and also some financial support from state and federal authorities. What was probably most staggering was his mobilisation of support among the local people: he led meetings attended by practically the entire adult population (political meetings are only attended by those who cannot avoid them: various local authorities, school children, people with political ambitions), organised voluntary work brigades and collected water fees which came to a considerable amount.[18] By the summer of 1971, the town had drinking water, and a group of adolescents had finished the first grade of secondary school without having to leave Tlayacapan.

Fabrizio's leadership emerged out of his access to, and manipulation of, a number of social and economic resources which were completely out of the reach of any local person (Varela, 1972). The implementation of public works necessitates the help of outside actors as the meagre budget of municipalities hardly covers anything but the low salaries of municipal officers and employees: and it is impossible to get financial support from outside without the mediation of a patron—such as Fabrizio. Inevitably, politics has to be situated in the context of patron–client relationships.[19]

Fabrizio's leadership did not last long, because his own actions provoked a series of contradictions at various levels. First, the group of Fabrizio's professional friends quarrelled among themselves and with him. The quarrel was related to ideological discrepancies: Humberto and Ordaz (the mathematics teacher) accused Fabrizio of being too authoritarian. They maintained that people should be more thoroughly consulted before determining a course of action. But ideological oppositions were reinforced by financial conflicts. Although both Humberto and Ordaz had small salaries from the school, and also some help from their families, they had become very dependent on Fabrizio's higher income as an architect, painter, sculptor and member of an extremely wealthy family. (They ate in Fabrizio's house, used his car, workshop, and even obtained money from him.)[20] The break with Fabrizio meant freedom for them, if perhaps also some economic stress. For Humberto, it also meant the possibility of finding a role in which he was not overshadowed by Fabrizio. The strategy followed by Humberto was to develop strong links with the *mayordomías*. Thus, he would obviously contrast with Mamerto—who disliked *mayordomos*—and with Fabrizio—who was not really interested in popular religion. Humberto founded a 'club' or meeting place for *mayordomos* in the chapel of Altica, in the Santa Ana *campan*, which was attended only by the *mayordomos* of that *campan*. He also started using radical, left-wing language in his sermons, which he ingeniously combined with symbols of popular religion, and published a weekly mimeographed pamphlet which he christened *El Pueblo* (which can be translated both as 'the town' and 'the people'). Both in the pamphlet and in his sermons he criticised the authoritarianism of Fabrizio—at first in a veiled manner, then quite openly.

The second contradiction concerned local political rivalries. The municipal council was altogether displaced by Fabrizio's overwhelming popularity. Another disaffected group was the primary teachers—they were no longer considered the local oracles, as the secondary teachers were now much more prestigious. Local power became a matter of being close to Fabrizio: it was his friends who were in a more favourable position to get benefits such as jobs in the well works, or simply prestige.

Thirdly, politicians at a higher level became worried about Fabrizio's power. There were rumours in Yautepec and Cuernavaca about Tlayacapan's being invaded by a group of foreigners. (In fact, Fabrizio belongs to a family of European immigrants, and the lawyer and his wife are Spanish-born.) Later, a state deputy openly told Fabrizio that 'they' (the governor and other state and federal authorities) could not allow local political offices to be allocated by a priest. Fabrizio's answer was that it had never been his intention to allocate offices to anybody. But his obvious influence could not be denied.

I shall briefly summarise the chain of events which followed the drilling of the well and which reveal the consequences of these three contradictions.

Fabrizio vested the title to the well in the Society of Users, of which he was the chairman and technical adviser. For the society's board of directors he chose the people he trusted, though he also appointed the members of the municipal council *ex officio* as consultants. The administration of the money—i.e. the task of collecting water fees and taking the money to the bank—he gave to the local public works committee. This committee had always existed and never had much importance (it organised people for tasks such as *jagüey* cleansing); it happened to be composed of five people whom Fabrizio thought could be trusted. It soon became clear that the only posts in the society that really mattered were the chairman and the treasurer. The municipal council grew sullen and jealous. Meanwhile, Fabrizio thought that some steps should be taken towards repaying the bank loan; so far the money collected had been enough to pay the interest, but not much of the capital. His first proposal was to sell water to a tourist project in the nearby lowlands. His second was to build public baths, showers and wash basins which people could use for a small fee. Then, he thought that something should be done to improve the financial circumstances of the townspeople. He founded a savings society, a consumer's co-operative to retail groceries at cheaper prices, a sewing shop for girls—for which he and his friends found customers in Mexico City—and a local building association. The latter was to take advantage of the growth of tourism in Tlayacapan, and hence a demand for weekend houses, made possible by the existence of the well. Fabrizio started building houses for tourists with a group of local masons—most of whom were trained by Fabrizio himself. To defend himself against accusations of authoritarianism, Fabrizio discussed his plans in public meetings, but his status was too high for anyone to oppose him.

At the end of 1973, a new municipal council was appointed. Its members were all friends of Fabrizio—connected to him through the savings society, the school parents' association, the building team, etc., or through having daughters in the shop. The council did not include a teacher. This precipitated the town into a major crisis. A campaign against Fabrizio and the council was openly instigated by Humberto—in his pamphlets—and by a teacher called Cutberto, who had been a member of the previous municipal council. Fabrizio was accused of promoting tourism to snatch land and water away from the town, of building houses for the rich, of exploiting local people for his personal businesses, and of misappropriating public funds. Fabrizio's followers and friends accused Humberto and the teacher of envy. Accusations were even made about the private lives of members of both parties.

Fabrizio's strategy was to defend himself publicly, and also to mobilise support from outside. His friend, the bishop, summoned Humberto and tried to placate him. After several interviews, Humberto wrote a public letter denouncing the bishop as a friend of the rich and an enemy of the people and their beliefs. Under pressure, Humberto left Tlayacapan and asked to be transferred to another diocese—although he came back by night at least once a week, to meet his *mayordomos*, who continued publishing the weekly pamphlet. But Cutberto, the teacher, could not be removed so easily. He asked the government that the finances of the well's Society of Users should be audited. Fabrizio consented provided it was done by official representatives of the state government. The state government sent its auditors, who declared that everything was in order. Then Cutberto denied the validity of this verdict, and organised a group which forcibly occupied the well premises. The municipal council tried to intervene; Cutberto then led a mob who took over the town hall. Fabrizio personally visited the state governor and the Minister of the Interior. He even got the daughter of the President of the Republic interested in the case. Everyone assured him that the law was on his side; but finally he was advised that the municipal council should resign so that a new election could take place. Reluctantly, the council resigned. The new election was won by Fabrizio's enemies. The well passed into the jurisdiction of a federal water committee, controlled by the Cuernavaca delegate of the Ministry of Hydraulic Resources. In 1976, Fabrizio left Tlayacapan.

Comments. The cleavages opened or deepened by Fabrizio's actions are related to a number of social factors which already existed in the Morelos Highlands. His quarrel with Humberto cannot be isolated from the uncertainty of the priestly role in the region, which is even more difficult when one priest is overwhelmingly overshadowed by the actions of another. Humberto felt it necessary to oppose and even to attack Fabrizio's achievements although this meant standing against his bishop. His support of traditional religion—so unusual among the Morelos priests—had the consequence of enhancing a personal prestige which otherwise he could hardly have acquired within the locality.[21]

The cleavages between local politicians resemble the fights in Atlatlahucan: for a local faction to survive, it must have members of the established political clique. In Atlatlahucan, Alfredo, an old politician, was removed; in Tlayacapan, it was Cutberto, the influential teacher, who resented Fabrizio's aloofness from him. The row among politicians combined with various other oppositions to develop into a feud between large sections of the town. Cutberto found support among many people of the Santa Ana *campan* where Humberto lived and had built a close-knit network through his *mayordomo* club; at one point, Humberto's and Cutberto's interests converged against Fabrizio, who lived in Exaltación. The Santa Ana people felt some resentment towards Exaltación: when Ambrosio had tried to gain the Lord of Exaltación *mayordomía* he was rejected and insulted because he belonged to Santa Ana. Most of the people who followed Cutberto in seizing the town hall were from Santa Ana, and they were very angry because the bishop had ordered Father Humberto to leave. The people of Santiago and Rosario were not so clearly defined in their allegiance. They supported or rejected Fabrizio according to their involvement in relationships of trust with people who worked for the architect-priest. Finally, people such as the moneylenders and storekeepers hated Fabrizio because of his consumer co-operative and savings association, although these enterprises never were strong enough to replace the local facilities.

The cleavage between Fabrizio as a local powerful leader and the external political authorities is related to the need of the Mexican political system to maintain fragmented power and multiple domains at the local level so that a unitary domain may emerge at the national level. There is, however, another external force which was hostile to Fabrizio: the tourist developers. Their projects have proliferated in the lowlands, and some *ejidos* have even been expropriated for building development. Fabrizio was visited by the representative of a big private building corporation who offered him money 'for his charitable enterprises' if he would persuade the local authorities to 'sell' parts of the *ejido* and the communal land. Fabrizio excused himself saying that he had no influence. Later, he was convinced that the developers had financed the campaign against him. In fact, Fabrizio was repeatedly attacked by Cuernavaca and Mexico City newspapers—which, as is commonly known, sell their space to the best buyer.[22] This space was too expensive for either Humberto or Cutberto. On the other hand, several important state and federal politicians hold shares in tourist development projects.

I have referred to all these cleavages as contradictions because there is no way of bridging them within the present system. In fact, the system causes them to exist—the emergence of important collective alliances which are co-terminous with the local setting are discouraged by centralised power. There was no compromise between Fabrizio and any of his enemies. Yet ... the local 'frames of reference' for ritual participation continue. I went to see the carnival in 1975, half expecting that it would fail. It did not. The *comparsas* of Santa Ana and Rosario-Exaltación attacked each other in their respective verses. But the colourful anonymity of the *chinelo* dance merged everybody into a single celebration, reminding Tlayacapaners that, after all, contradiction had always been a part of their daily competition for scarce resources.

The fragmentation of local power

A salient feature of *municipio* political life is the multiplicity of agencies concerned with maintenance of order, administration of civil matters and implementation of public goals. There is the municipal council, the *ejido* committee, the communal land committee, the village *ayudantes* (auxiliary officers), the public works committee, the party (whose delegates sometimes directly interfere in decisions). There is also local interference by the district police, the district court, the district deputy, etc. There is the tax collector—who is independent of the municipal council. All these agencies compete with one another for the control of relevant local environments; and it is only through the dependence on patrons in higher political spheres that a partial (never a total) control over environments is achieved. To demonstrate this, I first describe the operation of the municipal council, with special reference to data from Tlayacapan. I then show how the municipal council competes with other agencies for control over resources.

The functions of municipal councils. The municipal president is elected together with his two main assistants by the inhabitants of the *municipio* every three years. The two assistants are the *síndico* (syndic), and the *regidor de hacienda* (officer of the local treasury). These three officers constitute what is called the *ayuntamiento*, or municipal council. In addition, there are several employees of the *ayuntamiento*, who are directly appointed by the municipal president. These are the secretary, two local police constables, the *juez menor* (justice of the peace), and the tax collector in the open market.

The main function of the municipal president is to act as civil registrar for births, deaths and marriages. In this capacity he is directly helped by the secretary, who writes and types the certificates. The secretary is also in charge of filling in questionnaires on population statistics which are sent each year by the state and federal governments; he also keeps a register of cattle marks and sales, and is in charge of answering all official letters. The municipal president should oversee local morality according to the spirit of the national constitution. In this capacity he is assisted by the syndic—who functions as municipal attorney and who is in turn assisted by the justice of the peace and by the two police constables. In practice, this function involves (*a*) punishment of local minor offences, (*b*) arbitration in minor local disputes. Major offences and unsettled disputes have to be referred to the district court in Yautepec.

Minor offences frequently involve drunken behaviour; for example, urinating in public or accidentally breaking a window. In the evenings, especially on Saturdays, the constables tour the bars and send the drunkards home; if they resist—they rarely do—the constables may use physical force and put them in the local jail for the night. In cases of public misbehaviour, the municipal court—formed by the municipal president, the syndic and the justice of the peace—can impose fines up to 500 *pesos*, or, in the case of poor labourers, up to one week's salary. If a person does not appear in court after three summonses, the president can send him to jail for a maximum period of thirty-six hours, on grounds of contempt of court. If a person refuses to pay a fine, he can be jailed for a maximum period of two weeks. In practice, the *ayuntamiento* very seldom exercises its authority to punish. In 1970, for instance, the total sum of fines imposed on people in the town of Tlayacapan was 150 *pesos*, which were collected from four people accused of minor drunken offences.

The municipal authorities are also in charge of enforcing law and order in the other villages of the *municipio*. However, to a large extent each village functions independently; it has an *ayudante* elected by the village members every year, who registers births and deaths, and also fulfils the role of police. I will describe a case concerning the relationships between village and town authorities, and also some internal disputes within the town of Tlayacapan.

Case of gravel hill and the officer for communal land.

As already noted, *ejido* land is administrated separately from communal land. Communal land is mostly non-agricultural; it includes pasture, shrubs and barren land. Every village also has an officer for communal land. The limits of the communal land in each village are not very clearly defined; but this usually does not cause disputes as most of this land is not considered to be valuable. However, a dispute broke out between the municipal authorities of Tlayacapan and the village of San José over communal ownership of a hill situated between the town and the village.

The hill contains sand and gravel quarries. Nobody had paid attention to it until, in 1970, people from other parts of the region started to come to collect gravel and sand; people even came from outside the region. The villagers of San José knew that the hill belonged to them; however, they did not dare interfere. But in the summer of 1970 a man from San José, a teacher who lives outside the region, was visiting his parents. He thought that the village was being robbed; so he persuaded the villagers to guard the hill and exact a fee from every lorry wishing to collect gravel. At this, the lorry drivers complained to the Tlayacapan municipal president. The president thought that the fee should be paid directly to the *ayuntamiento*; but the villagers of San José disagreed and, led by the teacher, decided to resort to the Federal Department of Agrarian Affairs. [23]

The *ayuntamiento* had been elected very recently—it took office in June 1970. The president thought that the best way to deal with the matter would be to control the coming election of the *ayudante* of San José. He tried to appoint the new *ayudante* directly; but the villagers of San José refused to accept the president's man and elected someone else, who activated the legal procedures at the Federal Department of Agrarian Affairs. An officer of this department came and confirmed that the hill belonged to the villagers of San José and that they were entitled to collect fees from the lorries.

The municipal president of Tlayacapan then decided to survey the town's communal land to avoid similar situations occurring in the future. The man who currently held the post of communal officer had allegedly neglected his duties and failed to survey the land under his jurisdiction. So the president decided to convene an assembly for citizens of the whole town in order to elect a new officer, who would work in conjunction with the *ayuntamiento*.

The assembly elected a new officer. But the previous officer—called Eufemio—did not go to this meeting, and did not accept his dismissal. He claimed that his appointment did not depend on the municipal authorities, but directly on the Federal Department of Agrarian Affairs. Eufemio also claimed that the office should be elected only by the *comuneros*, i.e. by the people holding communal plots, and not by a general town assembly.

The *ayuntamiento* replied that the communal land was owned by the town as a whole, and demanded that all the relevant documents be given to the *ayuntamiento*. Eufemio refused; he was summoned three times by the president; as he did not appear, the constables put him in jail. Eufemio kept refusing to hand over any papers; after thirty-six hours, the *ayuntamiento* had to release him, and to resort to the Federal Department of Agrarian Affairs.

The department ruled that Eufemio's period of office was already over; but that the election should be made only by the *comuneros*. However, the list of *comuneros* was in Eufemio's possession, so the *ayuntamiento* could not call an election meeting, as no one knew who they were. All the *ayuntamiento* could do was to appoint the man elected by the assembly to make a new list so the meeting could be called; after a year and a half, this list was still incomplete.

Municipality versus district court. The disputes just described concern conflicts over authority at the *municipio* level. Other disputes concern competition between local and district authorities.

The function of the district court in Yautepec is to deal with major offences, and to settle the disputes which are not resolved in the municipal court. The district court has its own police—the *policía judicial*—which is entitled to interfere only in those cases where prosecution is directly made by the district court. However, the chief of the *judicial* police and one of his assistants come at weekends to the Tlayacapan bars, and collect machetes from the customers.To recover the machetes, the men have to travel to Yautepec and pay a fine of 20 or 30 *pesos*—not to the court, but to the police chief himself. On one occasion, a man who resisted being searched was beaten up by the *judicial* police. The local constables arrived, removed their guns, and expelled them. The case went to the state court, which supported the Tlayacapan constables. The *judicial* police nevertheless continue to visit the town, but avoid the local authorities.

Locally, people regard the district court with suspicion. It is known that the district officers can be bribed, so the Tlayacapan people prefer to solve their disputes locally. The following case illustrates how the District court is known to proceed.

A problem of fingers

Ciro's bar is very popular in Tlayacapan because it opens very early in the morning. On Mondays, and even on Tuesdays, some men go there for a drink to get rid of the weekend hangover. Some of them continue drinking for the rest of the morning—until their money, or their credit, runs out. One Tuesday this was the case with Lino, a poor labourer with a reputation as a drunkard. Around 10 a.m., Lino felt hungry and went to buy some tortillas in Pedro's shop opposite the bar. Pedro asked him to leave because he was upsetting his customers. Lino has the habit of embracing women in a friendly fashion when he is drunk. On this occasion Pedro began to push him gently out of the shop. Lino was so drunk that he tumbled over the threshold; then he tried to regain his balance by grasping the hinge of the iron door. At the same time, he leaned against the door, catching his hand in the hinge. As a result, the top joint of his right thumb was partially severed.

Ciro witnessed the accident from his bar, and ran to call Lino's wife and daughter, who came immediately and took him to the doctor. The twenty-two year old daughter returned soon after, and accused Pedro of having deliberately hurt her father. At noon, the constables summoned Pedro to the municipal court for the Wednesday evening. Pedro looked very nervous; he is a shy youth, twenty years old, and had never been involved in any kind of dispute. He is the son of one of the big local storekeepers—one who never participates in *fiestas*—so his family is not liked by the local people; however Pedro himself is regarded as being quiet and harmless.

On Wednesday evening the syndic listened to and wrote down Pedro's account of what had happened. Ciro confirmed that Lino was very drunk that morning. Lino replied that Pedro should pay all medical expenses, plus compensation money, as he would not be able to work for two or three weeks. Pedro said that he was willing to pay half of the cost of medical expenses, but not because he felt guilty, but out of friendship. Lino refused this offer. The syndic then said that he did not have authority to rule on the matter, and, as the disputants had not reached agreement, the case should go to Yautepec. In his report to the district attorney, the syndic stressed that Lino was a habitual drunkard.

At noon on the Friday, Lino and Pedro met the district attorney in Yautepec. The attorney was dressed up in suit and tie, and asked to be addressed as '*licenciado*' (the title given in Mexico to university graduates). His first words were for Pedro: 'Look, old chap,' he said, 'this accusation is serious. We can lock you up for quite a while.' Pedro tried to deny his guilt, but the attorney would not let him speak, and repeated the phrase several times. He then asked Lino if he accepted Pedro's offer to pay half the medical expenses, and Lino's answer was again negative. The attorney then dismissed Lino, and told Pedro to come back the following Monday for a further hearing.

On the Monday, Pedro asked his brother Mateo to go with him. Mateo is thirty-five years old; he helps his father run the store and owns a lorry. When he encountered the attorney, he went straight to the point.

'Good morning, *licenciado*. I come with my little brother to ask you if the problem of Lino's finger could be arranged between us.'

The attorney expressed surprise: 'How could we?'

Mateo nodded, and stretched two fingers.

'Well,' said the attorney, 'but the matter is quite serious.'

Mateo added another finger.

The attorney smiled. 'Let's write the report,' he said. While he typed, Mateo counted 300 *pesos* and passed the money to the attorney under the table.

When Lino went to Yautepec the following day, he was told that Pedro had proved his innocence. Pedro, in turn, told the story to his friends and soon after everybody in Tlayacapan knew what had happened in the district court. 'It is a question of fingers,' commented Ciro. 'That's the language spoken in Yautepec.' [24]

What are the rules of the game?

The case of Lino and Pablo—which is rather trivial—can in a very important respect be compared with the cases of the well and the communal land quarry—where more substantial issues were at stake. In all the cases the intervention of external political actors leads to the ending of a dispute, although it cannot be said that the disputes were really *settled*: their causes are still there. (In the case of the Atlatlahucan disagreement, the intervention of the government to support two municipal authorities reinforced the existing contradictions.) If matters remain unresolved, local actors cannot break their dependent ties with external power. It is pertinent to note that in none of the cases did the local people think that the intervention of external authorities had much to do with reaching a solution according to the law. In the case of Lino and Pedro, the decisive factor was bribery; in the case of the gravel pit, the municipal president was convinced that if he had a personal connection in the higher agrarian spheres, the pit would be placed in the hands of the municipal council (and he was probably right). The rules of the game are thus perceived as being a function of personal protection by a patron. Sometimes this protection can be bought with money. On other occasions, it can be gained through loyalty.

Loyalty and the perception of their own powerlessness are related to the participation by the local authorities in a series of activities which further reinforce their subordination and dependence. One straightforward example is the support they give to the state tax collector—who sends all the money outside the *municipio*. If they do not put pressure on slow taxpayers, it is said that the Cuernavaca state and federal treasury office would be unwilling to provide funds for public works. Another example is elections. The incumbent authorities make sure (by crossing the ballot papers themselves) that the candidates appointed by the official party win by an overwhelming majority. Thus the state party representatives are pleased and willing to reciprocate with both personal favours and support for municipal petitions—roads, schools. The vertical support that was obviously received by people like Alfredo and Cutberto can be partially explained in terms of their previous participation in political patron–client networks. On the other hand, the existence at the municipal level of a set of political 'old timers' facilitates patronage. Old timers can be trusted, and know what it is all about. New people are less reliable. Of course, new people have to be recruited; but it is much better that they collaborate with the old timers. (Fabrizio's followers were often young and unknown in political circles).

Why should local people become involved in such a farce? The quest for power and leadership cannot be separated from the context of local economic life. I have described a context of choice characterised by the increasing need for cash, the existence of risk and uncertainty, the involvement of actors in a wide social field, and their adoption of a variety of strategies, including the establishment of local relationships of trust and the diversification of resources. Political participation can be converted into economic assets—as the case of Arturo Figueroa shows—and it comes almost naturally to a man who achieves leadership prestige as a ritual organiser. In fact, I have never met a municipal president or *ejido* officer who had not been *mayordomo* or *autor* at least once. (Of course, the opposite is not true: *mayordomos* do not automatically become politicians. There are other factors involved: one of them is attachment to patrons and political networks.) And there is also the expectation of gaining higher political posts. The case of Matías Polanco, an old *zapatista* who became district deputy and then municipal president of Cuernavaca—and had also the chance of becoming governor—is often quoted. What is not said is that, lately, local politicians have got no further than humble jobs in the state bureaucracy: tax collectors, bodyguards, drivers. This would be 'dissonant' with expectations of political careers. And, after all, to get a salary—any salary—is always desirable.

The lack of local political power is consistent with, and related to, lack of economic power. The dependence of local politicians on political patrons is comparable to the dependence of local farmers on refineries, banks, moneylenders, wholesalers, transporters, multinational technology peddlers, importer-exporters, etc. On the other hand, none of the political patrons, or external economic forces, has a complete hold over local political life, or over local economic life. They are limited to control over specific resources—never over the complete environment of a given unit. In turn, patrons and economic forces depend on the support given to them by others—further up the hierarchy. At any of these points, government representatives are present. (The refineries depend on government control of water allocation and sugar prices; wholesalers, on government price policies, transporters and exporters on government permits.) The organisation of national society is a pyramid of patron–client relationships which permit concentration of power at the top.

The complexity of this fabric determines a dual phenomenon: individual manoeuvering is quite possible because it is impossible to achieve structural change. It has often been said that the basis for government power in the countryside is the jural centralisation of land distribution as well as the manipulation of agrarian demands by the C.N.C. (Conferación Nacional Campesina) through which all *ejidatarios* and *comuneros*, willingly or unwillingly, join the official party (Warman, 1977). But the rural population are also subordinated through agricultural production, migration, civil administration and public force. A man can use his relationships of trust to obtain individual power as long as such relationships do not become horizontal alliances which lead to the emergence of an independent faction: and as long as his political network includes external patrons and 'old timers'. This system of patronage is the basis for the functioning of the official party at a variety of levels.

The P.R.I. distributes political offices among its three sectors: the Confederación de Trabajadores Mexicanos (C.T.M.), formed by industrial workers; the Confederación Nacional de Organizaciones Populares (C.N.O.P.), whose membership is recruited among bureaucrats, teachers and shopkeepers, and the C.N.C. In the Yautepec district there is an implicit agreement: the C.T.M. nominates the state deputy—who is therefore chosen from refinery workers—whereas municipal councils are spoils for the C.N.C. and the C.N.O.P. In the land of Zapata, it is politically convenient to have *ejidatarios* as municipal presidents (thus the P.R.I. boasted in the 1970 presidential campaign that *ayuntamientos* had been taken over by the agrarian revolution); but C.N.O.P. old timers (teachers and shopkeepers) occupy the other posts. The direct control over the Highlands' councils has been traditionally vested in the hands of informal groups of *ejidatarios* and teachers, necessarily connected to Cuernavaca or Mexico City patrons through local old timers. The members of these groups, on the other hand, frequently compete for scarce resources. The Atlatlahucan and Tlayacapan feuds should be interpreted not as mere disputes between clergymen and their fans, but as the result of cleavages and rivalries among factions and subfactions of the C.N.C. and the C.N.O.P. (Fabrizio's sympathisers were *ejidatarios* and excluded teachers from public offices, whereas Efraín was on more friendly terms with teachers.) But, in the long run, both sectors tend to balance each other: they have the common function of manipulating and inhibiting collective demands in exchange for personal favours.

Notes

1. By 'Church' I mean the hierarchical and institutional aspects of established Catholic religion; by 'representatives of the Church' I mean bishops and priests. These operational definitions refer to a level specifically different from the theological.

2. The reasons why a specific item is found in religious ideologies or behaviour are particularistic and historical. The logical use of a set of items has to be understood from a sociological point of view.

3. It is surprising that Central American anthropologists have paid little if any attention to the role of the church in the life of the communities they studied. The terms 'Church', 'priest' and 'clergy' do not even appear in the index of the massive volume on social anthropology in the *Handbook of Middle American Indians*.

4. In the colonial period, the clergy could use their alliances with the Indian population—which were built from, and manifested in, ritual events—as an asset in transactions with royal officers. In contemporary Malta, Boissevain (1965: 45ff.) has described a situation in which social life is effectively controlled by the parish priest thanks to his ability to manipulate his position in ritual.

5. See Quirk (1964: 64). Oscar Lewis (1951: 259 ff.) describes an example of this in Tepoztlán.

6. In a pastoral letter, a prominent Mexican bishop declared that the peasants should be educated before getting any land, and that the existence of *hacendados* was necessary, because they could provide 'an example to their neighbours'. (Quirk, 1964: 68–9; see also Carrasco, 1957: 21).

7. For instance, the constitution abolished all religious schools and monastic orders. Religious institutions were denied juridical personality (they cannot hold any property). The priests do not have the right to vote, or to occupy any political office. Their use of clerical garb is restricted to places of worship—which have become the property of the nation, and are supervised by the civil authorities. For a discussion, see Mecham (1966: 385 ff.) and Quirk (1964:66).

8. An index of the growing importance of the Church is the increase in the number of clergy. In 1935 there were less than 500 priests in the whole country. In 1949 the total number was 4,327. In 1969 it was 8,649 (See Mecham, 1969: 404; F. C. Turner, 1971: 184).

9. In the nineteenth century the Protestant preachers who visited rural communities met strong hostility from the clergy. In some cases they were beaten and even murdered by mobs incited by the priests (González, 1956c: 368).

10. See the interview given by Méndez Arceo to Luis Suárez in *Siempre*! 10 November, 1971. ('*Mi tarea la he centrado en el anuncio del evangelio a partir de las ceremonias de la comunidad cristiana, en las que he hincado el bisturí de las transformaciones sociorreligiosas'*).

11. Méndez Arceo also protected radical religious experiments, such as the famous psychoanalytic monastery' in Cuernavaca, and has harboured Ivan Illich and his CIDOC, for which Emile Pin has lectured periodically.

12. I was told by one of the local priests that most confessions are about dreams and worries—not about 'sins'. Totolapan is the exception: folk religion is weak there, and the influence of the priest is stronger.

13. A Cuautla newspaper reported the incident (*El Polígrafo*, 31 January 1967).

14. All religious buildings are the legal property of the nation.

15. See Amerlinck's study of Protestantism in the region of El Mezquital (1970).

16. 'Fright' (*espanto*) refers to a situation of psychic maladjustment; it is perhaps the most frequent local idiom to express insecurity. Most *espantados* are children; but adults occasionally 'get it'. The ceremony is important because it provides an

opportunity to establish *compadrazgo* and *padrinazgo.*

17. There is an analysis of the drilling process written by Roberto Varela (1972). I only mention some points which are relevant for the analysis of the role of the priests and the nature of local politics.

18. Brigades for public work have a long tradition. School building and *jagüey* cleansing are usually done willingly. Yet everybody said that Fabrizio's *faenas* (public work tasks) were the best supported.

19. The municipal budget is based on three types of taxes: commercial (stores and open market), cattle (registering and selling), civil registry. Fines are also appropriated by the municipality. All this is usually exiguous, especially in a rural *municipio* where commerce and cattle raising are unimportant. Taxes on land are much more substantial, but they are paid to the state and federal governments (Varela, 1972:ch. II).

20. The lawyer (who had family wealth) and the doctor (who was paid by his patients) did not have this problem, and did not quarrel with Fabrizio.

21. Lately Humberto has developed alliances with the anti-bishop Atlatlahucan groups, and also with the Mexican followers of Marcel Lefebvre, the conservative French bishop.

22. *El Correo del Sur*, an independent Cuernavaca newspaper, consistently published truthful information about the Tlayacapan case (for instance, the issue published on 17 March 1974).

23. Now Secretaría de la Reforma Agraria.

24. This story was witnessed and recorded by Salvador Nuñez.

IX
AFTERWORD

Rather than a cursory summary of the themes developed in the book, this last chapter provides a brief view of some of their implications for social science research in Mexico. In so doing, I hope to draw attention to a number of issues which deserve further examination, in relation to both the Morelos Highlands and other areas. I also hope that this discussion will clarify why I have chosen to combine several levels of analysis: historical and contemporary, macrosocial and microsocial. As in any scientific venture, the first and the last questions concern the nature of the models used in the research.

Model building is almost universally considered to be a key characteristic of scientific activity, in so far as this is concerned with analysis and not with description. Models do not describe reality: they abstract certain crucial elements and relations, with the purpose of manipulating the confusing multiplicity of the empirical world. Models are the means of knowing reality in a precise manner; for precise knowledge requires a standpoint. They also illuminate certain features of reality: a researcher may never be aware of elements and relations without seeing their reflection in an abstract model.

It is, however, possible that a model becomes a means of avoiding knowledge, an excuse for ignoring aspects of reality which cannot be fitted into conceptual constructs. This often happens when there is no ongoing dialectical relationship between abstract thought and empirical research. Such a relationship implies the use of explicit intermediate variables, concepts which relate parcels of particular data to the general categories of the model—including those data where the applicability of the model is not immediately apparent.[1]

Consider, for example, the models which view social change as a unilineal process. Functionalist or Marxist, these models stress the standardisation of social links within overarching institutions: 'modern' capitalist firms, 'modern' bureaucracies. Authors such as Godfrey and Monica Wilson (1968: 24 ff.) define the process in terms of an increase in the 'intensity' of outside links which weakens internal links—i.e. commitment to impersonal duties and distant actors entails the diminution of commitments to personalised behavioural expectations. Parsons (1964: 34 ff.) makes a similar point with his well-known distinction between 'modern' and 'traditional' social roles in terms of dichotomies of 'pattern variables': neutral/affective, universalistic/particularistic, specific/diffuse, performance/quality, collectivity oriented/self oriented. If one were to take these definitions seriously, it would be expected that involvement in cash-crop, 'modernised' agricultural production would entail the emergence of social relationships established on neutral, universalistic, etc., bases. However, I have tried to present situations in which it is precisely the involvement of farmers in complex capitalist production which demands commitments with *compadres*, kinsmen and neighbours, and the establishment of highly 'diffuse' behavioural expectations. Furthermore, historical research shows that the very existence of a Morelos Highlands 'peasantry' throughout the centuries has been a function of its participation in a commercially oriented plantation economy.

From the standpoint of the Marxist classical model, what characterises non-capitalist relationships (as opposed to those which emerge in a capitalist mode of production) is their highly differentiated quality: you cannot simply substitute one relationship for another, as they are linked to particular persons or statuses and serve multiple functions.[2] Capitalist relationships are not differentiated: their associated services have become commodities and can be easily replaced in the context of a market economy. When Marx developed this idea in *Das Kapital*, he was making abstractions of historical conditions and sketching out ideal model features. But the indiscriminate use of the term 'capitalism' may be misleading for analysing concrete conditions. Again, the data of the Morelos Highlands show that people take pains to differentiate their relationships through ritual, and that this differentiation becomes highly functional for their participation in a commercial economy. Furthermore, such diffentiated relationships emerge not only in the context of small localities, but also in reference to actors situated at various points on a large-scale social spectrum: *encomenderos*, missionaries and royal officers in the colonial period; *hacendados* in the nineteenth century; commercial entrepreneurs, agents of multinational corporations, clergymen and politicians in the contemporary period. In order to analyse this variety of relationships, the concept of 'articulation of modes of production' is sometimes used. But it is important not to reify the concept; not to assume that two fundamentally separated organisational frameworks are united by certain links. It would be more appropiate to speak of the functional co-existence of different types of relations of production within a single organisational–capitalist–framework (Long, 1977).

The models of 'peasant' societies or segments emphasise only the existence of non-capitalist relations among small-scale rural cultivators. This emphasis may also be excessive and misleading. A recent formulation (Warman, 1976: 305 ff.) of the 'peasant mode of production' holds that the basic organisational frameworks are 'the family' (*sic*) and the community.[3] The first is defined as a production–consumption unit which allocates economic roles to its members; the second, as a bounded collectivity whose members are related to one another by symmetrical exchanges and participation in communal goals—including production goals for collective subsistence. I do not dispute that 'the family'—I would prefer to use the term household, or clusters of households—provides a context within which *some* activities, like work on subsistence agriculture, are organised. But the full range of economic activities both in the past and in the present has obviously necessitated wider organisational frameworks—in the Morelos Highlands and in other regions of rural Mexico. Present-day involvement in cash-crop farming by smallholders leads to a situation in which the significance of domestic labour decreases. Those farmers who manage to obtain some surplus use their wealth to take their children out of agriculture and into formal schooling and urban jobs; whereas poor farmers and their offspring spend more time working as paid labourers than cultivating their family plots. (Thus, the family plot becomes a supplement to salary, and not vice versa.) On the other hand, domestic labour cannot be displaced in a context of scarcity, risk and uncertainty. However, the use of domestic labour is not the only productive strategy—not even the main one—in the dynamics of regional agriculture. Nor is domestic consumption the main purpose of productive strategies: in a context of increasing monetisation and dependence on urban markets, production is inevitably directed to obtaining money to spend on everyday items, as well as to invest in technology, education, migration and economic diversification.

As for the symmetrical exchanges within the bounded community, it has to be said that even in the depressed periods of the late nineteenth and early twentieth centuries the Morelos Highland settlements presented sufficient stratification to cast doubts on the utility of such an approach. This stratification is perhaps not very obvious if one looks only at land distribution—it is also important to explore differentiation in terms of access to other resources, including political offices, migration, commerce and patron–client external ties. Implementation of communal goals may exist as a rallying point in certain situations—the question is how and by whom are these communal goals defined. It seems that it is really the external power which has defined communal goals; from colonial tributes and labour control, to land distribution and contemporary tax collection.[4] As a political form, the rural community has implied a mechanism of domain; to ignore this simple fact was an error of liberal governments. Of course, one important meaning of *fiestas* and ritual has always been the symbolic expression of communal unity. But I have argued that ritual groups are ephemeral and situation–specific, and that ritual settings should be regarded as 'frames of reference' where individuals build egocentric networks—and not where corporate groups emerge.

The problem of corporateness is, I hope, sufficiently discussed in the body of the book, especially in Chapter III. Chapter IV shows the drastic limitations of community, and even regionally based, action in the context of the revolution and agrarian reform. The material presented in Chapter VIII illustrated the enormous difficulties faced by any attempt to undertake independent communal action. One reason for the enthusiastic defence of the corporate model among anthropologists is perhaps the influence of British anthropologists' writings on corporate lineages in African tribal society. But it is precisely the tribal character of this society—or the fact that colonial rule was a recent phenomenon and followed the pattern of 'indirect' government—which forbids a mechanical comparison with Mexico, or Latin America generally.[5] A crucial variable in the latter is the strong impact of 300 years (at least) of colonial rule and the continuing presence of institutions influencing all levels of the social fabric—from the macrosocial to the local community. In short, Latin American society cannot be understood without the historical force of the State, pervading economics and politics, religion and kinship, ethnicity and class.

A new type of model should be used for analysing rural Mexico. I do not pretend to build such a model, though I would like to see the material provided by this book as helpful in that direction. The model should single out a number of analytical variables which do not simply lead to a broad conception of 'the Mexican countryman', but which help to discern variation—variation which is a result not only of cultural or geographical peculiarities, but of articulation in wider social processes. If Mexican peasantry emerges as such as a structural consequence of capitalaccumulation, then one has to take into account historical and regional modalities of accumulation in order to understand the peasantry. The concept 'peasant' is not monothetic but polythetic: it does not define a constellation of traits or attributes which are necessarily found in all the empirical entities to which the term is applied. (This being so, the search for a real definition [6] of *the* peasantry is a useless task.) Some of the attributes alluded to in the concept are present in certain circumstances; other attributes are present in other circumstances. What is analytically pertinent is to isolate the conditions in which certain attributes or sets of attributes appear and persist.

The importance of regional research projects defining such analytical variables cannot be sufficiently emphasised. Within this type of project, I would suggest the following themes:

1. *The historical roots of regional integration*: I have contended that ecology, and regional arrangements and boundaries, emerge out of history: the long term processes of organisation of resources. During the colonial age, the presence of market-oriented economic enterprises was crucial for defining division of labour and labour-force flows. The different character of these enterprises determined the nature of occupational diversification, and the intensity and extension of labour flows: sugar plantations, mining centres, cattle *haciendas*, had different requirements for permanent/seasonal labour—and also different methods of supplying food to their workers; thus, they 'created' regions of different types.

This process of regional creation was made possible by the support of elaborate State apparatus: the royal bureaucracy and the Church, whose presence it is also necessary to document. However, these apparatuses had different impacts on different regions: in places like Morelos they carefully regulated labour flows and protected selected subsistence land so that an inter-regional equilibrium could be achieved. In other places—Jalisco, for instance—they facilitated the expansion of *haciendas* producing different commodities—including cereals—on practically all types of land. In Jalisco, these *haciendas* became vast organisations which could diversfy their production and move their labourers from one task to another according to seasonal and market demands. The result was a flexible, highly mobile labour force; in addition, given the lack of inalienable land, labourers could acquire or rent plots and become *rancheros* (González, 1972; 37ff.). Such a situation (almost impossible in the Morelos Highlands) was not rare in times of market crisis; it also occurred when the *hacienda* became immoderately large and preferred to give land to its own men rather than losing control over it.

The different responses of local populations to the strategies of both the economic enterprises and the State is another point to be researched. The study of power is relevant in this connection. Also, the study of the uses of ethnicity, of migration flows, of ritual and political offices within villages. Regional demography—natural growth and migration flows—should be an essential variable in understanding the organisation of resources—particularly labour. (One obvious weakness of my historical chapter is the lack of detailed data on demographic evolution and distrubution: did periods of population growth in the Highlands correlate with stages of plantation expansion? Was the increase or decrease of population in particular villages influenced by the plantations? Was the overall Morelos population growth of the eighteenth and nineteenth centuries evenly distributed between lowlands and Highlands?)[7]

2. *Changes in regional boundaries*: The regional setting defined during the colonial period has conditioned the types of change that have come later. Revolution, for instance, was in certain regions a response to encroachment on land which had been defined as communal during the colony. The *cristero* civil war of the 1920s, on the contrary, did not occur in Morelos, but in regions such as Jalisco, where the post-revolutionary *ejido* distribution was affecting land which had *not* been communal during the colony.[8]

The organisation of land and labour in the contemporary period might be better understood if compared with the colonial situation and then with the changes suffered during the century of agrarian turmoils. In this respect, one could ask a host of important questions concerning the use of labour during both the dry and rainy season. Concepts such as power and strategy again become relevant.

In the Morelos Highlands, the nineteenth-century rearrangements did not change the boundaries of the region, in the sense that the focus of economic power was still the sugar plantations. A relevant question for other regions is the extent to which boundaries remained relatively untouched. Some of the agrarian measures of the post-revolutionary government could be related to the reinforcement or disruption of regional boundaries—and some of their consequences could be predicted in such terms. In Morelos, land reform measures seem to have maintained the dry season/rainy season, lowlands/Highlands labour pattern, providing the refineries with means of indirect control over *ejido* produce (through control of water, credit and marketing). In Jalisco, at least in its southern part, the post-revolutionary government has fostered the monopoly of land use for products which are not labour-intensive, such as timber and mechanised sorghum: this has led to a fragmentation of regions and sub-regions and to radical changes in the occupational structure and the direction of labour flows.[9]

What makes it possible—and important—to compare the regional impact of colonial and post-revolutionary policies is that in both periods the State has a common feature: a power domain is established with the purpose of organising the economy at all levels.

3. *Monetisation and labour alternatives*: Even if regional boundaries and basic labour-flow patterns are maintained, important changes emerge when the impact of a centralised cash economy is met by strategies of economic diversification. Such impact was felt in some regions (Jalisco is again a relevant example) as early as 1900, when railway-transported, mass-produced industrial goods replaced local, small-scale manufactures. The Morelos Highlands experienced the main impact of industrial cash economy after 1940. A response common to many regions is that the use of local land and resources is increasingly determined by the market possibilities of different products. Also, migration and education become means of re-orienting not only the flow but also the character of labour. Again, State policies—communications improvements, allocation of industries—in different regions are a crucial variable. The demographic variable also plays an important part in so far as it may be both a cause and a result of economic diversification—but probably not irrespective of environmental pressures for obtaining cash.

It is important to document in each region the ways in which money has become absolutely indispensable; yet I would argue that the basic trend is not from a non-cash to a cash economy, but from a less centralised to a highly centralised industrial economy. Both in Morelos and in Jalisco the use of money became general from the colonial period onwards. It was necessary to have money to meet ordinary necessities: that was the whole point of undertaking dry-season labour migration. But the Highlands situation in the 1930s was one in which people needed more money than they were actually getting—because labour in the lowlands was scarce. Inevitably, obtaining money became a priority. Later, competition for dry-season employment continued; but in addition Highlanders were participating in a commercial context of disadvantageous exchange *vis-a-vis* the industrial urban centres. Now, even 'traditional' *fiestas* require cash for expensive items—like spirits and gunpowder—produced in cities.

4. *The organisation of small-scale cash-crop economy*: The growing of cash-crops in smallholdings is widespread in Mexico. But different types of crops have different implications. High cash investments—as for tomatoes—reinforce monetisation trends. Risk and uncertainty relate to the persistence of 'diffuse', 'differentiated' social relations. And labour requirements relate to demographic density and migration patterns. Labour-intensive crops such as those found in the Morelos Highlands should foster immigration into the region and also the return of labour migrants. In such circumstances migration is frequently a stategy for participation in the local cash context. At the same time, local agricultural growth is severely limited: enterprises do not expand much, money is invested in education for children. In contrast, the expansion of a low-labour crop such as sorghum in the south of Jalisco is related to the actual decrease of population in many *municipios*, and to the permanent settlement of migrants elsewhere.

5. *The structure of local commitments*: I have emphasised the categorical nature of kinship relationships in the Morelos Highlands, and their functionality as a social idiom for developing relationships of trust. One has the impression that Mexican kinship is more a category than a structure. Nutini (1968) maintains that corporate kinship groups do exist in Tlaxcala—though, to my mind, he never demonstrates their viability. In any case, more comparative research is needed on the role of kinship in the changing Mexican economy: in rural Jalisco the situation is quite

similar to that of Morelos. On the other hand, it is striking to find so many ritual associations in the Morelos Highlands, as compared with other parts of Mexico or even Latin America. In Jalisco the Church has managed to preserve its power, and religious celebrations do not express selective social alliances, or they do it to a lesser extent than in the Highlands. Again, there is a lack of comparative studies which view ritual in the context of a regional commercial economy.

The study of kinship and ritual alliances is also necessary to illuminate the nature of trust relationships. Trust emerges from a process of selective social interaction. The selection of interacting partners begins in childhood. A child develops his first ties of trust among his parents' circle of relatives, *compadres* and friends, as well as among neighbours; however, as time goes by, trust becomes a function of persistent and satisfactory reciprocity in transactions—not of kinship or neighbourhood norms. A man develops a network of trusted friends which includes relatives and non relatives, neighbours and previous strangers. The composition of this network reflects his occupational and migratory career, his farming strategies, his political activities. Kinship terms, *compadrazgo* ceremonies, *fiesta* and carnival participation are symbolic reinforcements of reciprocity. As symbols sanction both symmetrical and asymmetrical transactions, they also reinforce social inequality. But symbols also express contradictions—e.g. among actors involved in a ritual activity—and, in situations of strife, they may reinforce processes of disruption of trust and subversion of the standing order.

6. *The urban centres*: The existence of lowland urban commercial centres which organised the distribution of regional crops has already been mentioned, but never analysed, mainly because of lack of data. Generally, we know almost nothing about the role of regional small cities in the context of regional agrarian structures. One reason for lack of research on this subject is the overwhelming influence of Mexico City as the seat of bureaucratic control, industrial growth and commercial monopoly. Yet it is often the intermediate urban groups which articulate the region to the centralised power. This is less true of Morelos—the sheer proximity of Mexico City permits the emergence of direct links—than of a place like Jalisco, where Guadalajara—the big city—and Ciudad Guzmán or Tepatitlán, the intermediate cities—organise migration flows and commercial exchanges to a large extent.

7. *The participation of individuals*: I have stressed the need to

combine historical and broad statistical sources with detailed observation of particular behaviour, and suggested the importance of focusing on the active participation of individuals and groups in the events and processes of regional change. Studies of local commitments could probably gain insight not by focusing on institutions—'the household', 'the family', 'the *mayordomía*'—but by adopting an actor-oriented approach. The 'convertibility' of resources of several kinds—including social resources—into assets within different spheres of value becomes a key concept for understanding the logic of life careers. Similarly, the significance of social links, cultural symbols and sanctions becomes more apparent when they are seen as people manipulate them to fit their aims. The old anthropological problem of cultural integration can be met in this way: culture only becomes integrated through human action. It is thus misleading to interpret the persistence of *fiestas*, for example, with a 'plural society' approach. But, in addition, the actor-oriented perspective permits one to explore more fully and deeply the effective social field of individuals and groups: 'to integrate both micro and macro level of analysis, and to give equal attention to horizontal and vertical patterns of co-operation and control' (Long, 1977: 189–90). Examining strategies at the micro level could be decisive for the analysis of power: actors are constrained by it in all their moves. Power is not simply a structure expressed in abstract terms by academics: it is a concrete dimension in individual biographies.

8. *The purposes of power and the problems of development.* In Mexico—as everywhere—the function of the State is the maintenance of the existing order, i.e. the maintenance of social inequality. A task for the social scientist is to document and analyse the regional strategies of the State, through the study of both the federal government and the actors who receive delegated power from it. (Different regional strategies reflect both differences among regions and the nature of the State itself.) It seems important further to explore the relationship between domain fragmentation at the local level and the emergence of a unitary domain at the national level,[10] in order to understand the limits to horizontal alliances, the necessary rivalry and competition at the local level, and the inevitable dependence on patrons connected to central power. On the other hand, the State apparatus is also in charge of distributing differential benefits among the population. The logic of this distribution transcends rural aspects and regional boundaries and refers to the historical dynamics of the Mexican nation, which manifests itself in the

configuration of the State at different periods of time and is conditioned by the existing correlation among international forces. But from such macro-political panorama one must descend to its implications for regional development.

One should be careful not to repeat clichés about development and under-development. But research must take into account the fact that regional poverty and limits to growth are a function of national distribution of power. In Morelos—and in Jalisco—agriculture is already 'developed' yet poverty persists. In the Morelos Highlands, agricultural problems are not merely endogenous: they originate in the relationships of producers with the U.S.A. market, the middlemen, the wholesalers. It is not lack of technique but lack of power, not lack of rationality (local actors are perfectly rational given the context of small-scale production) but lack of alternatives that makes local economy stagnant. Moreover, individual success in agriculture is related to participation in non-agricultural activities—including politicking—and this helps to prevent the emergence of collective responses to structural problems.

I would like to end this book with a sonorous sentence: a prophecy of social change, a bold statement on the inevitable consequences of unsolvable contradictions. But I am afraid of simply adding anthropological promises to the long list of political—and unfulfilled—promises made to Morelos Highlanders. Instead of easy recipes for 'better' societies, a commitment to critical knowledge is the only possible answer. 'As the other Latin American countries, Mexico must find its own modernity. In a sense, modernity has to be invented from the ways of living and dying, of producing and spending, of working and enjoying created by our people. It is a task demanding both favourable social and historical conditions and extraordinary realism and imagination. I do not need to recall that the re-birth of imagination—in art as in politics—has always been prepared and preceded by analysis and critique. Our generation and the one that follows have to assume this task. And, in the first place, we have to cure our intoxication from simplistic and simplifying ideologies' (Paz, 1978:44).

Notes

1. Knowledge is not a mere aseptic relationship between mind and object. In so far as the knowing subject has a social position and interests, his knowledge is conditioned by its possible or actual instrumentality. However, science demands that all assumptions and categories be constantly questioned through empirical research.

2. Gluckman—in another context—speaks of 'simplex' and 'multiplex' links; in the latter, 'nearly every social relationship serves many interests' (1955: 18–19). Multiplex relations tend to be long-living, stable and strongly binding, as divisions and conflicts emerging in one area of behaviour are balanced by necessary alliances in another. On the other hand, each particular simplex relationship may be ephemeral as long as it can be replaced without disrupting other social areas, or without losing access to crucial resources.

3. Warman uses Chayanov's ideas somewhat mechanically to interpret Mexican reality.

4. This point has also been made in reference to Peruvian communities. See Roberts and Samaniego (1976).

5. Indirect rule was created by Spain in colonial America. But the contrast with colonial Africa is that, in the latter, the pre-colonial organisation of local society managed to survive for a certain period of time. However, it was disrupted when colonial regimes demanded that labour force and resources become systematically incorporated into new economic enterprises.

6. Real definitions require monothetic terms; they strive to state 'the precise nature of a thing' (*Concise Oxford Dictionary*). As a social science normally utilises concepts which are directly taken from natural languages, real definitions in social analysis are rather scarce.

7. The data for answering all these questions probably could be found in the files of the Cuernavaca diocese, the Mexico City diocese, and in the General National Archive.

8. I shall make some tentative comparisons between Morelos and Jalisco—the two areas in Mexico in which I have done fieldwork (De la Peña *et al.*, 1977, also Fábregas, 1977).

9. In Jalisco, the economic diversification inherent in *hacienda* strategies resulted in high population density. In addition, *haciendas* also created interdependence among human groups and ecological areas. Such interdependence disappeared; nowadays, regional fragmentation is a result of vertical systems of organisation of resources, controlled by Mexico City groups.

10. Emergence is not the same as consolidation. A unitary domain cannot consolidate in a capitalist country, or wherever a minimum of democracy exists. However, one can speak of a trend—full of contradictions—towards federal concentration of power, in so far as federal agencies and tentacles become superordinated to state and municipal governments through tax and budget control and in so far as they tend to control those resources which are necessary for private enterprises, through legal manipulation and allocation of public expenditure.

APPENDIX I
A note on data collection

As I was interested in a wide range of data, I resorted to a variety of secondary material and to material collected at first hand by several techniques.

Chapter II is based on official census data and on direct observation. The latter plus preliminary historical research convinced me of the necessity of looking at a whole region—which I was able to circumscribe both as a geographically distinct zone and as a historical unity—and not at a single village or town. I lived in the town of Tlayacapan, but I made countless visits to all other *pueblos* and walked around the fields, hills and forests.

Chapter III is based on (*a*) a handful of unpublished documents from the National Archives of Mexico City (Archivo General de la Nación), (*b*) published collections of documents (Del Paso y Troncoso, 1905, 1939–42; Zavala and Castelo, 1939–45; Chávez-Orozco, 1953), (*c*) published works on colonial Central Mexico which throw light on and/or include Morelos data (Zavala and Miranda, 1954; Gibson, 1967; Coy, 1971; Gerhard, 1970, 1975; Carrasco, 1961, 1976a, 1976b), (*d*) monographs on Morelos (García-Martínez, 1969; Sotelo-Inclán, 1970; Barrett and Schwartz, 1975; Riley, 1973; Hernández-Orive, 1973), (*e*) social-historical literature on nineteenth-century Mexico (González 1956a, b and c). John Womack's 1969 monograph on Zapata deserves a special mention: its reading is essential for any perception of Morelos's past.

For Chapter IV I was able to consult all the documents in the Agrarian Reform Ministry files (Archivo de la Secretaría de la Reforma Agraria) relating to the Morelos Highlands *pueblos*. In addition, I consulted Tlayacapan local municipal files, and local *ejido* files in Tlayacapan, Totolapan and Nepopoalco.

Chapters V to VIII were written on the basis of direct observation, open interviews, questionnaires and structured interviews applied to a sample, and census material. The inadequacy of census data for anthropological analysis, as well as the particular inaccuracy of some Mexican censuses, has been often pointed out. Unfortunately, there is no other way of getting general information of a certain kind, unless one decides to conduct one's own census—which is only possible for recent information. Census data is one of several indicators of social reality, and has to be checked against information of other kinds. Questionnaires and structured interviews were carried out in the autumn of 1971 with a random sample of eighty Tlayacapan household heads (about 15 per cent of the total number of households in the town). This is referred to in the text as the 'Tlayacapan sample'. The town was chosen because it has been the most important settlement in the region for at least 300 years, and its population presents a range of activities wide enough to give an idea of variation and change. Open interviews with local people were carried out by twelve students working under my close supervision (five for one month, seven for two months) in the summer of 1971. We were able to select informants from the towns of Tlayacapan, Totolapan, Atlatlahucan and Tlalnepantla, and the villages of San José de los Laureles, Nepopoalco and San Agustín. The themes of these interviews included: local history; family structure and household economy; agricultural organisation; changes in crops and technology; local politics; ritual; education; migration patterns. My own observations were made while I was living in Tlayacapan from November 1970 to October 1971; on weekend visits during 1974, and during a summer vacation in 1975.

Finally, an important source of data and ideas were other anthropologists who have worked in Morelos: Robert Redfield, Oscar Lewis, Roberto Varela, Guillermo Bonfil, Teresa Rojas, John Ingham, Andrés Fábregas, Judith Friedlander and, particularly, Arturo Warman and his team.

APPENDIX II
Additional tables

Table 1

English equivalents of units of measurement

A. *Distance*
1 kilometre (km) = 1,000 metres (m) = 3,280 ft
1 metre = 100 centimetres (cm) = 1,000 millimetres (mm)
= 39.37 inches
1 mile = 1.6 km = 1,600 m

B. *Surface*
1 hectare (ha) = 10,000 square metres (m^2) = 2.471 acres

C. *Weight*
1 kilogram (kg) = 2.205 lbs
1 metrical ton = 1,000 kg

D. *Volume*
1 litre (lt) = $1\frac{3}{4}$ pints (approx.)

E. *Money* (1971)
1 *peso* = 3.5 new pence
1 pound stirling = 31 *pesos* (approx.)

F. *Temperature*
In order to convert Centigrade into Farenheit, multiply by 9/5 and add 32.
Example:

$$20\text{ C} = \frac{(20)(9)}{5} + 32 = 68\text{ F}$$

Table 2

Protein consumption in the Morelos Highlands

	Municipios			
Percentage of population consuming the following items twice a week and less	*Atlatlahucan*	*Tlalnepantla*	*Tlayacapan*	*Totolapan*
Meat	83	77	38	80
Eggs	71	69	44	68
Milk	83	93	62	85
Fish	99	98	99	99

Source: 1970 national census

Table 3
Distribution of private land in three communities

Size of holdings (in hectares)	*Number of owners*				
	Tlayacapan	*Totolapan*	*Nepopoalco*	*Sub-total*	%
Less than 1	139	141	30	310	36.5
1 – 2.5	108	114	39	261	31
2.6 – 5	82	30	27	139	16.4
5.1 – 10	47	15	36	98	11.5
More than 10	30	7	4	41	4.6
Total	406	307	136	849	100

Source: Municipal files

Table 4
Absolute and functional illiteracy in the Morelos Highlands

	Atlatlahucan		*Tlalnepantla*		*Tlayacapan*		*Totolapan*	
Age group	*A*	*B*	*A*	*B*	*A*	*B*	*A*	*B*
15–19	15	38	17	46	12	34	15	44
20–29	18	56	21	52	20	51	24	67
30–39	26	73	30	88	29	73	23	77
40 and over	38	80	47	88	43	82	42	87
Total	26	61	32	72	28	64	29	72

A. Percentage of absolute illiterates (declare to be so)

B. Percentage of functional illiterates (less than four years schooling)

Based on the 1970 national census

APPENDIX III

Angel Rojas's childhood and adolescence

I was born in 1922. My family was very poor, as was almost everybody else in the Morelos Highlands. My father had to hire himself out as a labourer. His daily salary was 75 cents; but he did not manage to get work every day. He also sowed chillies in his own field: chillies had a good market. We used to get up at five o'clock in the morning and go to the field, my brothers and I, to keep my father company. When I was six, my parents sent me to Doña Leovigilda's school, which opened at half past eight, so I had to leave the field at eight and run to classes. The school fee was 25 cents a month. We had no books—most of us could not afford them—but we learned a lot. At least I did. I can read, write and handle numbers very well. There were about fifteen of us at school. We all had nicknames. Mine was Diablo (Devil), because Doña Leovigilda had a child whose name was Angel, the same as me; but this Angel was ill and very quiet. The teacher christened me Diablo to distinguish me from her son.

In school, all children dressed the same. Boys wore *cotón* (long shirt) and *calzones* (pyjama type trousers) made of white *manta* (coarse cotton cloth). Girls wore *manta* blouses and skirts. All clothes were made at home by our mothers. One could get ready-made clothing at Yautepec and Cuautla; occasionally some Turks came to the Highlands villages with clothing; but it was very expensive for us. Cheap *manta* could be found in the weekly *tianguis* (open market), or in Yautepec. I remember that my brother Alfredo went to work as a porter to Mexico City, and came back dressed in an American jacket and trousers. His clothes were everyone's envy.

I suppose I was rather obnoxious as a child. My nickname was not totally undeserved. Doña Leovigilda's husband kept some sheep and goats in the backyard—and he used to cook delicious barbecues for the whole school on the teacher's birthday; it was heaven for us, because meat was a very rare dish in those days. Anyhow, Angel, the sick child, had a pitcher full of goat milk for himself every day. Sometimes I got very hungry and thirsty, and I went to the kitchen and drank the milk. (The school was in the front room of the teacher's house, so it was easy to slip into the kitchen when she was not looking.) I was caught several times, and Doña Leovigilda beat me with a stick. I used to get beaten once or twice a week for all sorts of things. The teacher's fawners were in charge of bringing her sticks periodically from a tree nearby. But Don Lauro, the teacher's husband, liked me a lot. He was a good violinist—he even organised a band; I used to mount one of the goats and 'dance' with it while he played the violin and laughed heartily. There was a big water tank in the backyard—very clean water for drinking; several times, when the weather was very hot, I took my clothes off and jumped into the tank. Finally the teacher found out because my hair was wet, the water surprisingly low, and there was water everywhere. So I said: 'The turkey fell into the tank and splashed water over the backyard', but she was unconvinced. At the time, a muleteer charged 3 cents for a shipment of *jagüey* water. My father had to pay 3 *pesos*, a hundred shipments, to refill the tank, after I had cleaned it out.

Of course, I was also beaten by my father (not with a stick but with a *barzón*—the leather thong which is used to tie the plough) and by the priest. I had such a bad reputation that whenever something bad happened the blame was put on me. Yet sometimes, not very often, I was innocent. I must say that Doña Leovigilda never beat me without reason. Probably my biggest feat was to hunt a *zorrillo*—one of those black-and-white foxes whose urine smells like hell. A *zorrillo* was hiding in the baker's house, in the chimney; so the baker said: 'I shall give you one *peso* and a bag of bread if you get that beast out.' I gained the prize but the *zorrillo* left his mark on me. People could smell me several blocks away. When I arrived home, my father would not let me in—he did not even want to look at the bread.

My brother Dámaso and I started working when he was fourteen and I was twelve, harvesting maize and beans. Both of us had left school already. The first week we earned 5 *pesos* between the two of us, which seemed a lot of money. After the harvest, there was no work to do. We were very young to go and work elsewhere, so the only thing was to help our father. I found an additional occupation: stealing poultry. I did it with a group of friends—who were as hungry as I was—but I was the expert. We lured the chicken out of the yard with maize, and then, far away in the hills, cooked it in a casserole, and ate it. To this day, I can cook delicious chicken casseroles. I must have killed 300 chickens. Once I was reported to the authorities, and had to pay for the stolen birds. My parents were ashamed of me, so I stopped robbing backyards. Then I killed ducks. Many ducks came to the swamps, channels and dams in Oaxtepec, and they were also good to eat.

At fourteen I discovered bulls. I had always liked the bull rings which were a part of San Juan's *fiesta*, but had never mounted a bull before. The first time I did it, my mother beat me with a big stick. She thought it very dangerous. Yet I was so fascinated with the whole thing that I could not stop. Once my mother saw that I was putting my spurs on, and sent for the police constable. 'This boy is under eighteen,' she said. 'He cannot do dangerous things without my permission.' So the constable locked me in jail for the whole afternoon, until the bull ring had finished. I started going behind my parents' back. *Fiesta* organisers in other places invited me to their bull rings. Thus I became famous everywhere: in Casasano, Calderón, Oaxtepec, Oacalco, Totolapan, Tepoztlán, Nepopoalco . . . I even went to Milpa Alta several times with friends who took their cattle over the Ajusco, to sell in the fair. Then I found out that some of the cattle these fellows took were not their own. They were *abigeos*, cattle thieves, so I stopped going with them. This was fortunate, because they were executed a few months later. It was the custom around here to hang *abigeos* on the spot. Anyhow, I travelled all over the place without paying a cent, for people wanted me to ride in *fiestas*. I never fell from a bull. I had many girl friends, and many people gave me food and drinks. In Calderón, I was even offered an *ejido* plot by the authorities: they wanted me to stay there. I really had a good time. But I reckon that the danger was real, because our bull rings are very risky. One has to stay on the bull without saddle or stirrups. The only help is the *pretal* (a leather thong tied around the animal's chest). One has to grip the *pretal* with one hand. The other hand is used to salute one's friends with a sombrero. The bull runs and jumps from the very moment he feels the weight of a man on his back; in addition other bull fighters provoke the bull with their red capes, so he runs like mad from one corner to another. Then the bull shakes his head against the capes. This is most dangerous because the rider may be hurt by a horn if he is not good enough to keep his head straight and high. I always could avoid the horns, except once. The *pretal* slipped—it was not sufficiently tight—I leaned—and lost my left eye. I was lucky. Other riders have died in similar circumstances.

Just after losing my eye, I had a big revolutionary adventure. The *presidente municipal* announced that all young men over eighteen had to do military service, because our country was getting ready for the war against the Germans. We were very angry at this news. Our families needed our work, and we wanted no business with the Germans, nor with the *gringos*, who were apparently on our side. There was a famous rebel at the time, a man called Encarnación Barreto, who had a hideout in the eastern mountains of Morelos. He said he was Zapata's heir and that he was going to drive the Federal Army from our land. However, people in the Highlands did not pay much attention to Barreto. Nobody wanted another decade of hunger and killing. But Barreto used the discontent over the army draft to win supporters. One night, a friend of mine came to see me and invited me to meet Barreto's envoy. The meeting was at midnight, in a ravine. A man who called himself 'the captain' said that we had to fight for our freedom; otherwise, we would spend perhaps ten years in the army, fighting for a cause we did not even understand. Five youths of Tlayacapan agreed to join the movement, and followed the captain at once. I was one. He said that a great popular army was waiting for us in the mountains to take Cuautla. In fact, we met only three other people. None of them seemed to know what was going on, but they were armed. We weren't. After two days' wandering without apparent purpose in the mountains, with no food except one or two old *tortillas* which the 'captain' had in his pocket, we decided to quit. 'This is nonsense,' I said. 'We'd better go back to Tlayacapan and join you later, when your popular army is ready.' The 'captain' seemed to be fed up as well, but he threatened to kill us. His fellows dissuaded him; they argued that the shots might attract attention, and that Barreto's reputation would suffer if his officers started killing youngsters. So we went back to Tlayacapan, hoping that our absence had not been noticed.

But it had. Soon after I got home, the *presidente municipal* came to visit me and politely said that I was expeted in Cuernavaca on the following day. So I went to Cuernavaca, lest the government took measures against my family. I was surprised to see that more than 100 youths from the Highlands were also going. A truck collected us on the new dirt road, just outside Tlayacapan.

In Cuernavaca, we were taken to a big army house, with a beautiful patio and many trees. There was a big meal, with many soldiers and officers, and we were invited to it. During the meal, a general spoke. 'Look here, young men,' he said, 'the army is not as bad as they say. We are training Mexican citizens, not to go to war, but to be better men. We teach them to read and write, we teach them discipline and work. Besides, not all of you are going to be drafted, only those chosen in the lottery. From all of you, our fatherland wants peacefulness and production.' And we went back to Tlayacapan on the same day. I was never drafted. Encarnación Barreto was found and killed less than one year after my revolutionary experience.

REFERENCES

A.G.N.:Archivo General de la Nación.

A.I.N.A.H.: Archivo del Instituto Nacional de Antropología e Historia.

A.M.T.: Archivo Municipal de Tlayacapan.

A.S.R.A.: Archivo de la Secretaría de la Reforma Agraria.

Adams, Richard N. (1965a), 'Introduction', in *Contemporary Cultures and Societies in Latin America*, edited by Dwight B. Heath and Richard N. Adams, New York: Random House, pp. 4–16.

Adams, Richard N. (1965b), 'Social organisation: introduction' in *Contemporary Cultures and Societies in Latin America*, edited by Dwight B. Heath and Richard N. Adams, New York, Random House, pp. 257–87.

Adams, Richard N. (1970), *Crucifixion by Power: Essays on Guatamalan National Social Structure, 1914–1966,* Austin and London: University of Texas Press.

Aguilar Mora, Manuel, and Jorge Medina (1972), 'Sinaloa a la hora de los latifundistas', *La Cultura en Mexico*, no. 543.

Aguirre-Beltrán, Gonzalo (1952), 'Problemas de la Población Indígena de la Cuenca del Tepalcatepec', *Memorias del Instituto Nacional Indigenista,* vol. 3.

Aguirre-Beltrán, Gonzalo (1953), *Formas de Gobierno Indígena*, Mexico City; Imprenta Universitaría.

Aguirre-Beltrán, Gonzalo (1967), *Regiones de Refugio*, Mexico City: Instituto Indigenista Interamericano.

Aguirre-Beltrán, Gonzalo (1970), *El Proceso de Aculturación en Mexico*, 2nd Edition, Mexico City: Universidad Iberoamericana.

Aguirre-Beltrán, Gonzalo, and Ricardo Pozas (1954), 'Instituciones indígenas en el Mexico actual', in *Métodos y Resultados de la Política Indigenista en México*, by Alfonso Caso *et al*, Mexico City: I.N.I. (*Memorias del I.N.I.*, vol. 6, pp. 171–268).

Alejo, Francisco Javier (1970), *Estrategia de desarrollo económico de México: 1920–1920,* Unpublished thesis, Mexico City: Escuela Nacional de Economía,

U.N.A.M.

Althusser, Louis (1973), *Aparatos Ideológicos de Estado*, Mexico City: Escuela Nacional de Antropología e Historia, Ediciones de la Sociedad de Alumnos.

Amerlinck, Mari-Jose (1970), *Ixmiquilpan: Un Estudio Comparativo de Evangelistas y Católicos*, M.A. Thesis, Universidad Iberoamerica (Mexico City).

Amin, Samir (1975), *Categorías y Leyes Fundamentala del Capitalismo,* Mexico City: Editorial Nuestro Tiempo.

Arias, Patricia, and Lucía Bazán (1977), CIVAC: un Proceso de Industrializacion en una Zona Campesina, Mexico City: Centro de Investigaciones Superiore del I.N.A.H.

Amas, Patricia, and Lucía Bazán (1979), *Demandas y Conflicto. El Poden Politico—en un Pueblo de Morelos*, Mexico City: Nueva Imagen/C.I.S.—I.N.A.H.

Azaola, Elena (1976), 'Tepalcingo: la dependencia politica de un municipio de Morelos', in *Los Campesinos de la Tierra de Zapata*, by A. Warman et al, Mexico City: S.E.P.—I.N.A.H.

Bailey, F. G. (1957), *Caste and the Economic Frontier: A Village in Highland Orissa,* Manchester: Manchester University Press.

Bailey, F. G. (1960), *Caste, Tribe and Nation*, Manchester: Manchester University Press.

Baran, Paul (1957), *The Political Economy of Growth,* New York: Monthly Review Press.

Barlow, Robert H. (1949), 'The extent of the empire of the Culhua-Mexica' *Ibero Americana,* 28, Berkeley and Los Angeles: University of California Press.

Barrett, Ward J., and Stuart B. Schwartz (1975), 'Comparación entre dos economías azucareras coloniales: Morelos, Mexico, y Bahía, Brasil', in *Haciendas, Latifundios y Plantaciones en America Latina*, edited by Enrique Florescano, Mexico City: Siglo XXI.

Barth, Fredrik (1963), 'Introduction', in *The Role of the Entrepreneur in Social Change in Northern Norway,* Bergen/Oslo/Tromso: Universitets-forlalget, Scandinavian University Books.

Bartra, Roger (1974) *Estructura Agraria y Classes Sociales en Mexico*, Mexico City; Editorial Era (Serie Popular).

Bartra, Roger (1975) 'Campesinado y Poder político en México', in *Caciquismo y Poder político en el México Rural*, by Roger Bartra *et al.*, Mexico City: Siglo XXI and Instituto de Investigaciones Sociales de la U.N.A.M. pp. 5–30.

Beals, Ralph (1967), 'Acculturation', in *Handbook of Middle American Indians*, vol. 6: *Social Anthropology*, edited by Manning Nash, Austin: University of Texas Press, pp. 53–86.

Benítez Zenteno, Raul (1972), 'Política de población', in *El perfil de Mexico en 1980*, by Jorge Basurto *et al.*, vol. 3. Mexico City: Siglo XXI, pp. 551–80.

Boeke, J. H. (1953), *Economics and Economic Policy of Dual Societies*, Haarlem: H. D. Tjeenk Willink.

Boissevain, Jeremy (1965), *Saints and Fireworks: Religion and Politics in Rural Malta*, London School of Economics, Monographs on Social Anthropology no. 30, London: Athlone Press.

Bonfil, Guillermo (1971), 'Introducción al ciclo de ferias de cuaresma en la región de Cuautla, Morelos (Mexico)', *Anales de Antropología*, Instituto de Investigaciones

Historicas, U.N.A.M., pp. 167–202.

Bonfil, Guillermo (1973), *Cholula: La Ciudad Sagrada en la Era Industrial*, Mexico City: Universidad Nacional Autónoma de México.

Brading, David (1975), *Mineros y Comerciantes en el México de los Borbones*, Mexico City: Fondo de Cultura Económica.

Breedlove, James M. (1966), 'Effect of the Cortes, 1810–1822, on Church reform in Spain and Mexico', in *Mexico and the Spanish Courts*, edited by Nettie Lee Benson, Austin and London: University of Texas Press, pp. 112–33.

Camara, Fernando (1952), 'Religious and political organization', in *Heritage of conquest: the ethnology of Middle America*, edited by Sol Tax, Glencoe: The Free Press, pp. 142–73.

Campos, Margarita, (1973), *Escuela y Comunidad en Tepetlaoxtoc*, Mexico City: Colección Sepsetentas.

Cancian, Frank (1965), *Economics and Prestige in a Maya Community: The Religious Cargo System in Zinacantan*, Stanford: Stanford University Press.

Cancian, Frank (1972), *Change and Uncertainty in a Peasant Economy*, Stanford: Stanford University Press.

Carrasco, Pedro (1957), 'Tarascan folk religion' in *Synoptic Studies of Mexican Culture*, New Orleans: Middle American Research Institute, Tulane University, publ.17.

Carrasco, Pedro (1961), 'The Civil-religious hierarchy in Mesoamerican communities: pre-Spanish background and colonial development', *American Anthropologist*, vol. 63, pp. 483–97.

Carrasco, Pedro (1964), 'Family structure of sixteenth-century Tepoztlán', in *Process and Pattern in Culture: Essay in Honor of Julian H. Steward*, edited by Robert A. Manners, Chicago: Aldine, pp. 185–210.

Carrasco, Pedro (1976a), 'Estratificacion social indigena en Morelos durante el siglo XVI', in *Estratificación social en la Mesoamérica prehispánica*, by Pedro Carrasco, Johanna Broda *et al.*, S.E.P.-I.N.A.H.

Carrasco, Pedro (1976b), 'La sociedad mexicana antes de la conquista', in *Historia General de Mexico*, vol. 1. Mexico City, El Colegio de México, pp. 165–288.

Centro de Investigaciones Agrarias (1973), *Estuctura Agraria y Estuctura Agrícola en Mexico*, Mexico City: Fondo de Cultura Economica.

Cepes (1969), *Morelos: 5° Distrito: Yautepec* (Statistical Report), Cuernavaca: Cepes.

Chávez-Orozco, Luis (ed.) (1953), *Indice del Ramo de Indios del Archivo General de la Nación*, Mexico City: Instituto Indigenista Interamericano.

Chayanov, A. V. (1966), *The Theory of Peasant Economy*, edited by Daniel Thorner, Basil Kerblay and R.E.F. Smith, Homewood, Illinois: Richard D. Irwin for the American Economic Association.

Cline, Howard (1952), 'Mexican community studies', *Hispanic American Historical Review*, vol. 32, pp. 212–42.

Cortés, Hernán (1975), *Cartas de Relación*, Mexico City: Porrúa.

Coy, Peter (1971), 'A watershed in Mexican rural history: some thoughts on the reconciliation of conflicting interpretations', *Journal of Latin American Studies*, vol. 3, part 1, pp. 39–57.

Creel, Martha (1977), *Chiconcuac: Pueblo de Artesanos y Capitalistas*, Unpublished thesis, Mexico City: Departmento de Antropología Social, Universidad

Iberoamericana.

Cunniff, Roger L. (1966), 'Mexican municipal electoral reform', in *Mexico and the Spanish Cortes*, edited by Nettie Lee Benson, Austin and London: University of Texas Press, pp. 59–86.

Curiel, Andrés de (1905), 'Relación de Totolapan y los de su partido', in *Papeles de la Nueva España*, vol. 6, edited by Francisco del Paso y Troncoso, Madrid, pp. 6–11.

Cyert, R. M., and J. G. March (1970), *A Behavioural Theory of the Firm*, Englewood Cliffs, New Jersey: Prentice-Hall.

Dalton, George (1971), 'Introduction', in *Economic Development and Social Change: The Modernization of Village Communities*, edited by George Dalton, New York: American Museum Sourcebooks in Anthropology, The Natural History Press.

Dalton, George (1972), 'Peasantries in anthropology and history', *Current Anthropology*, vol. 13, nos. 3–4, pp. 385–415.

De la Peña, Guillermo, *et al.* (1977), *Ensayos Sobre el Sur de Jalisco*, Mexico City: Centro de Investigaciones Superiores del Instituto Nacional de Antropologia e Historia (Cuadermos de la Casa Chata, 4).

De la Peña, Guillermo (1977), 'Economic Development and social change in the region of the Morelos Highlands', University of Manchester Ph.D. Thesis.

De la Peña, Guillermo (1980), *Herederos de promesas. Agricultura, politia y ritual en los Altos de M* (1916), Mexico City: Ediciones de la Casa Chata.

De la Peña, Joaquim, José Crowley, Santos Amaro, and Hector Calles (1955), *Notas Sobre la Industria Azucarera de Mexico*, Mexico City: Edicion y Distribucion Ibero Americana de Publicaciones.

Del Paso y Troncoso, Francisco (ed.) (1905), *Papeles de la Nueva España*, Madrid.

Del Paso y Troncoso, Franciso (ed.) (1939–42), *Epistolario de la Nueva España 1505–1818*, 16 volumes, Mexico City: Biblioteca Histórica Mexicana de Obras Inéditas.

Del Villar, Samuel I. (1976), 'Depresión en la industria azucarera mexicana', in *Las Fronteras del Control del Estado Mexicano*, by Rafael Segovia *et al.* Mexico City: El Colegio de México.

De Moraes Filho, Evaristo (1970), 'Sociología del desarrollo de América Latina', in *Sociología del Desarrollo Latinoamericano (Una guía para estudio)*, by Pablo González Casanova *et al.* Mexico City: U.N.A.M.

Díaz del Castillo, Bernal (1974), *Historia de la Conquista de la Nueva España*, Mexico City: Porrúa.

Díaz, Mary N. (1966), *Tonalá: conservatism, responsibility and authority in a Mexican town*, Berkeley and Los Angeles: University of California Press.

Durkheim, Emile (1960), *De la Division du Travail Social*, 7th edition, Paris: Presses Universitaires de France.

Durkheim, Emile (1968), *Les Regles de la Methode Sociologique*, 17th edition, Paris: Presses Universitaires de France.

Epstein, A. L. (1958), *Politics in an Urban African Community*, Manchester: Manchester University Press.

Esteva, Gustavo (1975), 'La agricultura en Mexico, de 1950 a 1975: el fracaso de una falsa analogía', *Comercio Exterior*, XXV, 12.

Evans-Pritchard, Edward Everett (1937), *Witchcraft, Oracles and Magic Among the Azande*, Oxford: Clarendon Press.

Evans-Pritchard, Edward Everett (1940), *The Nuer*, Oxford: Clarendon Press.

Fábregas, Andrés (1969), *El Nahualismo y su Expresión en la Región de Chalco Amecameca*, Unpublished thesis presented at the Escuela Nacional de Antropología e Historia, Mexico City.

Fábregas, Andrés (1977), 'La formación histórica de los Altos de Jalisco', *Controversia*, vol. 1, no. 4.

Fallers, Lloyd A. (1961), 'Are African cultivators to be called peasants?', *Current Anthropology*, vol. 2, pp. 108–10.

Favier, Claudio (1977), *Ruinas de una Utopía San Juan Tlayacapan*, Ph.D. Thesis, University of Madrid.

Festinger, Leon, *et al.* (1964), *When Prophecy Fails*, New York: Harper Torch Books.

Flores, Edmundo (1970), *Vieja Revolución, Nuevos Problemas.* Mexico City: Cuadernos de Joaquin Mortiz.

Flores Magón, Ricardo (1970), *Antología*, edited with an Introduction by Gonzalo Aguirre Beltrán, Mexico City: Universidad Nacional Autónoma de México, Biblioteca del Estudiante Universitario, no. 93.

Florescano, Enrique (1969), *Precios del Maíz y Crisis Agrícolas en México (1708–1910)*, Mexico City: El Colegio de Mexico.

Florescano, Enrique (1971), *Estructuras y Problemas Agrarios de México (1500–1821)*, Mexico City: Sepsetentas.

Florescano, Enrique, and Isabel Gil (1976), 'La época de las reformas borbónicas y el crecimiento económico: 1750–1808', in *Historia General de Mexico*, vol. 2. Mexico City: El Colegio de Mexico, pp. 183–302.

Foster, George M. (1960–61), 'Interpersonal relations in peasant society', *Human Organization*, vol. 19, pp. 174–8.

Foster, George M. (1967a), 'Introduction: What is a peasant?', in *Peasant Society: A Reader*, edited by May Díaz; Jack Potter and George M. Foster, Boston: Little, Brown and Company.

Foster, George M. (1967b), *Tzintzunzan: Mexican Peasants in a Changing World*, Boston: Little, Brown and Company.

Foster, George M. (1969), 'Godparents and social networks in Tzintzuntzan', *Southwestern Journal of Anthropology*, vol. 25, no. 3.

Frank, André Gunder (1970), *Latin America: Underdevelopment or Revolution*, New York and London: Monthly Review Press.

Frank, André Gunder (1971), *Capitalism and Underdevelopment in Latin America*, Harmondsworth: Penguin Books.

Friedlander, Judith (1975), *Being Indian in Hueyapan*, New York: St. Martin's Press.

Furnivall, J. S. (1948), *Colonial Policy and Practice: A Comparative Study*, Cambridge: Cambridge University Press.

Furtado, Celso (1970), *Economic Development of Latin America*, Cambridge: Cambridge University Press.

Garbett, Kingsley, and Bruce Kapferer (1970), 'Theoretical orientations in the study of labour migration', *New Atlantis*, vol. 2, no. 1, pp. 179–97.

García, Antonio (1967), *Reforma Agraria y Economía Empresarial en América Latina*, Santiago de Chile: Editorial Universitaria.

García Cantú, Gastón (1971), 'Constitución o capitalismo', *Excelsior*, 3 September.

García-Martínez, Bernardo (1969), *El Marquesado del Valle: Tres Siglos de Régimen Señorial en Nueva España*, Mexico City: El Colegio de México.

Garza-Sánchez Máximo, *et al.* (1968), *La Economia del Estado de Morelos*, Mexico City: Sistema Banco de Comercio.

Geertz, Clifford (1966), *Agricultural Involution: The Process of Ecological Change in Indonesia*, Berkeley and Los Angeles: University of California Press.

Gerhard, Peter (1970), 'A method of reconstructing precolumbian political boundaries in Central Mexico', *Journal de la Societé des Americanistes*, vol. 59, pi. 27–41.

Gerhard, Peter (1975), 'La evolución del pueblo rural mexicano 1519–1973', *Historia Mexicana,* vol. 24, no. 4, pp. 566–78.

Germani, Gino (1956), *Politica y Sociedad en una Epoca de Transicion e la Sociedad de Masas,* Buenes Aires: Paidos.

Gibson Charles (1952), *Tlaxcala in the Sixteenth Century,* New Haven: Yale University Press.

Gibson, Charles (1967) *Los Aztecas Bajo el Dominio Español (1519–1810),* translated into Spanish by Julieta Campos, Mexico City: Siglo XXI.

Gluckman, Max (1955), *Custom and Conflict in Africa,* Oxford: Basil Blackwell.

Gluckman, Max (1958), *Analysis of a Social Situation in Modern Zululand,* Manchester: Manchester University Press (Rhodes-Livingstone Papers, no. 28).

Gluckman, Max (1963), *Order and Rebellion in Tribal Africa,* London: Cohen and West.

Gluckman, Max (1965), *Politics, Law and Ritual in Tribal Society,* Oxford: Basil Blackwell.

Godelier, Maurice (1968), *Rationalité et irrationalité en économice,* Paris: Francois Maspero.

Godelier, Maurice (1975), 'Modes of production, kinship and demographic structures', in *Marxist Analysis and social anthropology*, edited by M. Bloch, London: Malaby Press.

Gomez Sahagun, Lucila (1971), *San Miguel Tlaixpan: su organización político Social,* Unpublished thesis, Mexico City: Departamento de Antropologia Social, Universidad Iberoamericana.

González, Casanova, Pablo (1962), 'Sociedad plural y desarrollo: el caso de México ', *America Latina,* Año 5, no. 4.

González Casanova, Pablo (1965), *La Democracia en México*, Mexico City: Editorial Era.

González Jácome, Alba, *et al.* (1977), *Industrializacion en Tlaxcala: Papers presented at the Industrial Anthropology Symposium,* Mexico City: Universidad Iberoamericana (mimeo).

González, Luis (1956a), 'El hombre y la tierra', in *Historia Moderna de Mexico, La Republica Restaurada*, vol. 2, *La Vida Social,* edited by Daniel Cosío Villegas, Mexico City: Editorial Hermes.

González, Luis (1956b), 'El subsuelo indígena', in *Historia moderna de México*, vol. 2, *La vida social*, edited by Daniel Cosío Villegas.

González, Luis (1956c), 'Los campesinos', in *Historia Moderna de Mexico, La Republica Restaurada,* vol. 2, *La Vida Social,* edited by Daniel Cosio Villegas, Mexico City: Editorial Hermes.

González, Luis (1972), *Pueblo en Vilo*, Mexico City: El Colegio de Mexico.

Haag, J., and A. Soto (1969), *El Mercadeo de productos agropecuarios en México,* Mexico City: Limusa.

Hall, Basil (1924), *Extracts from Journal Written on the Coasts of Chile, Peru, and Mexico, in the years 1820, 1821, 1822,* 2 vols., Edinburgh, Printed for Archibald Constable and Co. (Edinburgh) and Hurst, Robinson and Co. (London).

Hansen, Roger (1971), *The Politics of Mexican Development,* Baltimore: Johns Hopkins Press.

Hernández-Orive, Alicia, (1973), *Haciendas y pueblos en el estado de Morelos: 1535–1810,* Unpublished thesis, El Colegio de México.

Hobsbawm, E. J. (1975), 'Vulnerable Japan', *New York Review of Books,* vol. 22, no. 12, pp. 27–32.

Holt-Buttner, Elizabeth (1962), 'Evolución de las localidades en el estado de Morelos segun los censos de población (1900–1950)', *Anuario de Geografía,* U.N.A.M.

Humboldt, Alejandro de (1953), *Ensayo político sobre el reino de la Nueva España,* Mexico City: Compañía General de Ediciones.

Huizer, Gerrit (1970), *La Lucha Campesina en México,* Mexico City: Centro de Investigaciones Agrarias.

Ingham, John M. (1968), *Culture and Personality in a Mexican Village,* Ph. D. Thesis, Berkeley: University of California (Ann Arbor, Michigan: University Microfilms Inc.).

Ingham, John M. (1970a), 'On Mexican folk medicine', *American Anthropologist,* vol. 72, no. 1, pp. 76–87.

Ingham, John M. (1970b) 'The asymmetrical implications of godparenthood in Tlayacapan', *Man,* vol 5, pp. 281–9.

Ingham, John M. (1971), 'Time and space in ancient Mexico: the symbolic dimensions of Clanship', *Man,* vol. 6, no. 4, pp. 615–29.

Kapferer, Bruce (1966), 'Review of M. G. Smith's *The Plural Society in the British West Indies', Rhodes-Livingstone Journal,* no. 38, pp. 62–4.

Kapferer, Bruce (1969), 'Norms and the manipulation of relationships in a work context', in *Social Networks in Urban Situations,* edited by J. Clyde Mitchell, Manchester: Manchester University Press.

Kapferer, Bruce (1972), *Strategy and Transaction in an African Factory,* Manchester: Manchester University Press.

Kautsky (1974), *La Cuestión Agraria,* Mexico City: Ediciones de Cultura Popular.

Kemper, Robert V. (1976), *Campesinos en la ciudad. Gente de Tzintzuntzan,* Mexico City: Sep Setentas.

Krober, A. L. (1948), *Anthropology,* New York: Harcourt Brace and Co.

Kunkel, J. H. (1965), 'Economic autonomy and social change in Mexican villages', in *Contemporary Cultures and Societies in Latin America,* edited by D. H. Heath and R. N. Adams, New York: Random House, pp. 438–53.

Labastida, Julio (1972), 'Los grupos dominantes frente a las alternativas de cambio', in *El perfil de Mexico en 1980,* by Jonge Basuto *et al.*, vol. 3, Mexico City: Siglo XXI, pp. 99–164.

Leach, Edmund (1954), *Political Systems in Highland Burma,* London: Athlone Press.

Leeds, Anthony (1965), 'Brazilian careers and social structure: a case history and model', in *Contemporary Cultures and Societies in Latin America,* edited by D. H. Heath and R. N. Adams, New York: Random House, pp. 379–404.

Lenin, V. I. (1950), *El desarrollo del Capitalismo en Rusia,* Moscow: Progreso.
Lenin, V. I. (1976) *Teoría de la Cuestion Agraria,* Mexico City: Ediciones de Cultura Popular.
Lewin, Kurt (1952), *Field Theory in Social Science,* London: Tavistock.
Lewis, Oscar (1951), *Life in A Mexican Village: Tepoztlan Restudied,* Urbana: University of Illinois Press.
Lewis, Oscar (1970), 'The folk-urban continuum', in *Urban Anthropology,* edited by Aidan Southall, Chicago: Aldine.
Lira, Andrés, and Luis Muro. (1976), 'El siglo de la integraciones' in *Historia General de Mexico,* vol. 2, Mexico City: El Colegio de Mexico, pp. 83–181.
Lombardo de Ruiz, Sonia (1973), *Desarrollo Urbano de Mexico—Tenochtitlan segun las fuentes historicas,* Mexico City: S.E.P.I.N.A.H.
Lomnitz, Larissa (1975), *Cómo sobreviven los marginados,* Mexico City: Siglo XXI.
Long, Norman (1968), *Social change and the individual,* Manchester: Manchester University Press.
Long, Norman (1970), 'Rural entrepreneurship and religious commitment in Zambia', *International Yearbook for the Sociology of Religion,* vol. 6, pp. 142–57.
Long, Norman (1972), 'Kinship and associational networks among transporters in rural Peru: the problem of the 'local' and the 'cosmopolitan' entrepreneur', Paper prepared for seminar series at the Institute of Latin American Studies, London University, entitled *Kinship and Social Networks in Latin America.*
Long, Norman (1975), 'Multiple enterprise in Highland Peru', Paper prepared for the Advanced Seminar on the Economic Anthropology of Investment Behaviour in Latin America, School of American Research, Santa Fe, New Mexico.
Long, Norman (1977), *An Introduction to the Sociology of Rural Development,* London: Tavistock.
Long, Norman and Bryan Roberts (eds) (1978), *Peasant Cooperation and Capitalist Expansion in Central Peru,* Austria: Institute of Latin American Studies, The University of Texas.
López, Sinesio (1973), 'Hueyapan: pueblo de la tierra fría', in *Los Campesinos de la Tierra de Zapata,* by Arturo Warman *et al.,* Mexico City: S.E.P.-I.N.A.H.
López-Beltrán, Lauro (1969), *Fray Antonio de la Roa, Taumaturgo y Penitente,* Mexico City; Ed. Jus.
Maccoby, Michael (1967), 'Love and authority: a study of Mexican villagers', in *Peasant Society: A Reader,* edited by Jack Potter, May Díaz and George Foster, Boston: Little Brown and Company, pp. 336–46.
Madsen, William (1967), 'Religious syncretism', in *Handbook of Middle American Indians,* vol. 6, Social Anthropology, edited by Manning Nash, Austin: University of Texas Press, pp.369–91.
Martínez-Marín, Carlos (1968), *Tetela del Volcán: su Historia y Su Convento,* Mexico: U.N.A.M.
Martinez, Tomas, and Leticia Gándara (1976), *Política y Sociedad en los Altos de Jalisco,* Mexico City: S.E.P.-I.N.A.H.
Marx, Karl and Friedrich Engels (1969), 'The German Ideology', in *Basic Writings on Politics and Philosophy,* edited with an Introduction by Lewis S. Feuer, London: Collins (Fontana Library), pp. 287–302.
McClelland, D.C. (1961), *The Achieving Society,* New York: Free Press.

Mecham, J. Lloyd (1966), *Church and State in Latin America,* revised edition, Chapel Hill: University of North Carolina.

Meillassoux, Claude (1977), *Mujeres, Graneros y Capitales*, Mexico City: Siglo XXI.

Mendizábal, Miguel Othón de (1974), 'El origen historico de neustras clases medias', in *Ensayos Sobre las Clases Sociales en Mexico*, by M. O. de Mendizabal, *et al.*, Mexico City: Nuestro Tiempo, pp. 9–22.

Mintz, S. W. and Eric R. Wolf (1950), 'An analysis of ritual co-parenthood (compadrazgo)', *Southwestern Journal of Anthropology,* vol. 6, No. 4. pp. 341–65.

Mitchell, James Clyde (1956), *The Kalela Dance,* Manchester: Manchester University Press (*Rhodes-Livingstone Journal*, no.27).

Müller, Florencia (1949), *Historia Antigua del Valle de Morelos,* Mexico City: Instituto Nacional de Antropologia e Historia.

Muñoz, Humberto, Orlandina de Oliveira, and Claudio Stern (1972), 'Migración y movilidad ocupacional en la ciudad de Mexico', in *El perfil de Mexico en 1980,* Vol. 3, Mexico City: Siglo XXI, pp. 325–58.

Murra, John V. (1975) *Formaciones Económicas y políticas del Mundo Andino*, Lima: Instituto de Estudios Peruanos.

Myrdal, Gunnar (1957), *Rich Lands and Poor,* New York: Harper and Row.

Nash, Manning (1967), 'Indian economies', in *Handbook of Middle American Indians,* vol. 6, *Social Anthropology,* edited by Manning Nash, Austin: University of Texas Press, pp. 87–102.

Nash, Manning (1968), 'The social context of economic choice in a small society', in *Economic Anthropology: Readings in Theory and analysis,* edited by Edward E. LeClair Jr. and Harold K. Schneider, New York: Holt, Rinehart and Winston pp. 311–22.

Navarrete, Ifigenia M. de (1970), 'La distribución del ingreso en México; Tendencias y perspectivas', in *El Perfil de México en 1980,* by David Ibarra *et al.*, vol. 1, Mexico City : Siglo XXI, pp. 15–71.

Neruda, Pablo (1970), *Canto General (I),* 4th edition, Buenos Aires: Losada.

Neyra, German (1976), *Zacualpan,* Unpublished thesis, Universidad Iberoamericana.

Nutini, Hugo (1965), 'Polygyny in a Tlaxcalan community', *Ethnology,* Vol. 4, no. 2, April, pp. 123–47.

Nutini, Hugo (1968), *San Bernardino Contla: The Social Structure of a Village in Tlaxcala,* Pittsburgh: Pittsburgh University Press.

Nutini, Hugo and Timothy D. Murphy (1970) 'Labour migration and family structure in the Tlaxcala-Pueblan area, Mexico', in *The Social Anthropology of Latin America,* edited by W. Goldschmitt and H. Hoijer, Berkeley and Los Angeles: University of California Press, pp. 80–103.

Ortiz, Sutti: (1971), 'Reflections on the concept of 'peasant cultures', and 'peasant cognitive systems', in *Peasants and peasant societies,* edited by T. Shanin, Harmondsworth: Penguin Books.

Palerm, Angel (1952), 'La civilización urbana', *Historia Mexicana,* vol. 2, no. 2, pp. 184–209.

Palerm, Angel (1967), 'Agricultural systems and food patterns', in *Handbook of Middle Americans Indians,* vol. 6, *Social Anthropology,* edited by Manning Nash, Austin: University of Texas Press.

Palerm, Angel (1968), *Productividad Agrícola: Un Estudio Sobre México,* Mexico

City: Centro Nacional de Productividad.

Palerm, Angel (1972) *Agricultura y Sociedad en Mesoamérica,* Mexico City: Sep Setentas.

Palerm, Angel, and Eric Wolf (1972), *Agricultura y Civilización en Mesoamérica,* Mexico City : Sep Setentas.

Palerm, Angel (1973), *Obras Hidraulicas Prehispánicas en el Sistema Lacustre del Valle de México,* Mexico City: S.E.P.-I.N.A.H.

Palerm, Angel (1974), 'Factores históricos de la clase media en Mexico', in *Ensayos sobre las clases sociales en México,* by M.O. De Mendizabal *et al.*, Mexico City: Nuestro Tiempo, pp. 91–108.

Palerm, Angel (1974), *Un modelo marxista para la formación socioeconómica colonial?* Mexico City: C.I.S.-I.N.A.H. (mimeo)

Palerm, Angel (1975), *El sistema Metropolitano y la articulación y Sucesión de Modos de Producción,* Mexico City: C.I.S.-I.N.A.H.

Parsons, Talcott (1951), *The Social System,* Glencoe: Free Press.

Paz, Octavio (1970), *Postdata,* Mexico City: Siglo XXI.

Paz, Octavio (1978), 'El ogro filantrópico', *Vuelta,* 21, pp. 38–44.

Pérez-Lizaur, Marisol (1975), *Poblacion y Sociedad: Cuatro Comunidades de Acolhuacan,* Mexico City: S.E.P.-I.N.A.H.

Piddocke, Stuart (1968), 'The Potlatch system of the southern Kwakiutl: a new perspective', in *Economic Anthropology: Readings in Theory and Analysis,* edited by Edward LeClair Jr and Harold K. Schneider, New York: Holt, Rinehart and Winston, pp. 283–99.

Pieris, Ralph (1969), *Studiues in the Sociology of Development,* Rotterdam: Rotterdam University Press.

Pin, Emile (1963), *Elementos Para una Sociología del Catolicismo Latinoamericano,* Madrid: Feres.

Pitt-Rivers, Julian (1965), 'Who are the Indians?', *Encounter,* vol. 25, no. 3., pp. 41–9.

Pitt-Rivers, Julian (1967), 'Words and deeds: the Ladinos of Chiapas', *Man,* N.S. vol. 2. no 1, pp. 71–86.

Quirk, Robert E. (1964), 'Religion and the Mexican social revolution', in *Religion, Revolution and Reform: New Forces for Change in Latin America,* edited by William V. D'Antonio and Frederick B. Pike, London: Burns and Oates.

Ramírez, Ramón (1973), 'La bola chiquita', in *Los Campesinos de la Tierra de Zapata,* by Arturo Warman *et al.*, vol 1. Mexico City: S.E.P.-I.N.A.H.

Rascón, Maria Antonieta, and Rafael Cordera,'Morelos a la hora del movimiento obrero independiente', *La Cultura en Mexico,* 1 November.

Redfield, Robert (1941), *The folk Culture of Yucatan,* Chicago: University of Chicago Press.

Redfield, Robert (1953), *The Primitive World and its Transformations,* Ithaca, N.Y.: Cornell University Press.

Redfield, Robert (1953), *The Little Community: Peasant Society and Culture,* Chicago: University of Chicago Press.

Redfield, Robert, (1964), *The Collected Papers of Robert Redfield,* 2 vols., Chicago: University of Chigago Press.

Redfield, Robert (1965), 'Culture changes in Yucatan', in *Contemporary Cultures and Societies in Latin America,* edited by D. H. Heath and R.N. Adams, New York:

Random House, pp.17–29.

Redfield, Robert, and Tax, Sol (1952), 'Introduction', in *Heritage of Conquest: The Ethnology of Middle America*, edited by Sol Tax, Chicago: University of Chicago Press.

Reyes Osorio, Sergio, and Solomon Eckstein (1974), 'El desarrollo polarizado de la agricultura mexicana', in *La Sociedad Mexicana: Presente y futuro*, edited by M. Wionczek, Mexico City: Fondo de Cultura Ecónomica, pp. 18–32.

Ricard, Robert (1947), *La Conquista Espiritual de Méxịco,* Mexico City: Editorial Jus.

Riley, G. Michael (1973), *Fernando Cortes and the Marquesado in Morelos*, Albuquerque: University of New Mexico Press.

Robelo, Cecilio (1887), *Diccionario de Nombres de Morelos* , Mexico City (n.p.).

Roberts, Bryan (1973), 'Migración urbana y cambio en la organización provincial de la sierra del Peru', *Ethnica*, no. 6.

Roberts, Bryan (1974), 'The interrelationships of city and provinces in Peru and Guatemala' in *Latin American Urban Research*,vol.4, edited by W. Cornelius and F. Trueblood, Beverly Hills : Sage Publications.

Roberts, Bryan, and Carlos Samaniego (1976), 'La reforma agraria en la siérra del Peru', *Problemas del Desarrollo,* vol 6, no. 25.

Rogers, Everett M. (1969), *Modernisation among peasants: The Impact of Communication,* New York: Holt, Rinehart and Winston.

Rojas, Teresa (1973), 'La cerámica conutemporánea de Tlayacapan, México', *Anales de Antropologia,* vol 10, pp. 241–64.

Rojas, Teresa (1974), 'Aspectos Tecnológicos de las obras hidráulicas coloniales', in *Nuevas Noticias Sobre las Obras Hidráulicas Prehispánicas y Coloniales en el Valle de México,* by T. Rojas, A. Strauss and J. Lameiras, Mexico City: S.E.P.—I.N.A.H., pp. 19–133.

Ronfeldt, David (1973), *Atencingo: The Politics of Agrarian Struggle in a Mexican Ejido*, Stanford: Stanford University Press.

Sáenz, César, (1975),'Xochicalco, Morelos', in *Los Pueblos y Señoríos Teocráticos,* by E. Matos *et al.* Mexico City: S.E.P.- I.N.A.H.

Sanders, William T. (1967), 'Settlement patterns, in *Handbook of Middle American Indians,* vol. 6, *Social Anthropology,* edited by Manning Nash, Austin: University of Texas Press.

Sanders, William T., and Barbara J. Price (1968), *Mesóamérica: The Evolution of a civilisation,* New York: Random House.

Semo, Enrique (1973), *Historia del Capitalismo en México,* vol 1. *Epoca colonial,* Mexico City: Era.

Sepulveda, Manola (1976), *La política Educativa en la Década de los Treinta: Las Escuelas Regionales Campesinas,* Unpublished thesis, Mexico City: Escuela Nacional de Antropolgía e Historia.

Shanin, Teodor (1971), 'Introduction' in *Peasants and Peasant Societies,* edited by T. Shanin, Harmondsworth: Penguin Books.

Silva Herzog, Jesús (1964), *El Agrarismo Mexicano y la Reforma Agraria,* 2nd edition, Mexico City: Fondo de Cultura Economica.

Simpson, Lesley Bird (1952), *Many Mexicos,* 3rd edition, Berkeley and Los Angeles: University of California Press.

Smelser, Neil J. (1967), 'Towards a theory of modernisation', *Tribal and Peasant Economies,* edited by G. Dalton, Garden City, N.Y.: Natural History Press, pp. 29–48.

Smith, M. G. (1965), *The Plural Society in the British West Indies,* Berkeley and Los Angeles: University of California Press.

Sotelo-Inclán, Jésus (1970), *Raíz y Razón de Zapata,* 2nd edition, Mexico: Comision Federal de Electricidad.

Stavenhagen, Rodolfo, *et al.* (1968), *Neolatifundismo, Explotación,* Mexico City: Nuestro Tiempo.

Stavenhagen, Rodolfo (1968), 'Seven fallacies about Latin America', in *Latin America, Reform or Revolution,* edited by James Petras and Maurice Zeitlin, Greenwich, Conn: Fawcet Premier, pp. 13–31.

Stavenhagen, Rodolfo (1969), 'Marginalidad y participación en la reforma agraria mexicana', *Revista Latinoamericana de Sociologia,* vol 5, no. 2. , pp. 249–74.

Steward, Julian (1956), *Theory of Culture Change,* Urbana: University of Illinois Press.

Strickon, Arnold (1965), 'Hacienda and plantation in Yucatán: historical-ecological construct of the folk-urban continuum', *América Indígena,* vol 25, pp. 320–45.

Suarez, Luis (1971), 'Mendez Arceo dice: me he ajustado al articulo 130', *Siempre!* 10 November, pp. 32–6.

Suárez, Luis (1972a), 'Agricultura en crisis', *Siempre!* no. 1003, 13 September.

Suárez, Luis (1972b), 'El mercado, Waterloo de la produción', *Siempre!* , no. 1001, 30 August.

Swartz, M. J. (1968) 'Introduction' in *Local-level Politics,* edited by M. J. Swartz, Chicago: Aldine Publishing Co.

Swartz, Marc J., Victor Turner and Arthur Tuden (eds.) (1966), *Political Anthropology,* Chicago: Aldine Publishing Co.

Tamayo, Jorge L. (1962A), *Geografía General de México,* 3 vols, 2nd edition, Mexico City: Instituto Mexicano de Investigaciones Economicas.

Taylor, William B. (1972), *Landlord and Peasant in Colonial Oaxaca,* Stanford: Stanford University Press.

Tello, Carlos (1968), *La Tenencia de la Tierra en México,* Mexico City: U. N. A. M.

Toledo, Francisco de (1924) [1570–74], 'Libro de la visita general del Virrey', *Revista Histórica* (Lima), vol. 7, pp. 115–216.

Trejo, Saúl (1970), 'El incremento de la producción y el empleo industriales en México', *Demografía y Economía,* vol. 4, no. 1.

Turner, Frederick C. (1971), *Catholicism and Political Development in Latin America,* Chapel Hill: University of North Carolina Press.

Turner, Victor W. (1974), *Dramas, Fields and Metaphors: Symbolic Action in Human Society,* Ithaca and London: Cornell University Press.

Van Velsen J. (1960), 'Labour migration as a positive factor in the continuity of Tonga tribal society', *Economic Development and Cultural Change,* vol. 8, pp. 265–78.

Varela, Roberto (1972), *Processus Politiques a Tlayacapan,* Paris: Ecole Pratique des Hautes Etudes (microfiches).

Villoro, Luis (1976), 'La revolución de independencia', *Historia General de Mexico,* vol. 2, Mexico City: El Colegio de Mexico, pp. 303–56.

Warman, Arturo (1972), *La Danza de Moros y Cristianos,* Mexico City: Sep Setentas.

Warman, Arturo (1973), *Los Campesinos, Hijos Predilectos dal régimen,* Mexico City: Nuestro Tiempo.

Warman, Arturo (1975), 'El neolatifundio mexicano: expansión y crisis de una nueva forma de dominio', *Comercio Exterior,* vol. 25, no. 12, pp. 1369–74.

Warman, Arturo (1976), . . . *y Venimos a Contradecir: Los Campesinos de Morelos y el Estado,* Mexico City: Ediciones de la Casa Chata.

Warman, Arturo *et al.* (1974–76), *Los Campesinos de la tierra de Zapata* 3 vol, Mexico City: S. E. P.-I. N. A. H.

Weber, Max (1946), 'Politics as a vocation', in *From Max Weber: Essays in Sociology,* edited by Hans Gerth and C. Wright Mills, New York: Oxford University Press.

Weber, Max (1974), *Historia Económica General,* Mexico City: Fondo de Cultura Económica.

Wharton, Clifford R. (1971), 'Risk, uncertainty, and the subsistence farmer', in *Economic Development and Social Change: The Modernisation of Village Communities,* edited by George Dalton, Garden City: Natural History Press, pp. 556–74.

Whetten, Nathan L. (1948), *Rural Mexico,* Chicago and London: University of Chicago Press.

Wilson, Godfrey, and Monica Wilson (1968), *The Analysis of Social Change,* 2nd edition Cambridge: Cambridge University Press.

Wolf, Eric (1953), 'La formación de la nación', *Ciencias Sociales,* vol. 4, pp. 50–62, 98–111, 146–71.

Wolf, Eric R. (1955), 'Types of Latin American Peasantry: a preliminary discussion', *American Anthropologist,* vol. 57, no. 3, pp. 452–71.

Wolf, Eric R. (1957), 'Closed corporate peasant communities in Mesoamerica and Central Java', *Southwestern Journal of Anthropology,* vol. 13, pp. 1–18.

Wolf, Eric R. (1959), *Sons of the Shaking Earth,* Chicago: University of Chicago Press.

Wolf, Eric R. (1965), 'Aspects of group relations in a complex society: Mexico', *Contemporary Cultures and Societies of Latin America,* edited by Dwight B. Heath and Richard N. Adams, New York: Random House, pp. 85–101.

Wolf, Eric R. (1966), *Peasants,* Englewood Cliffs, New Jersey: Prentice-Hall.

Wolf, Eric RX (1967), 'Sociocultural levels of integration', in *Handbook of Middle American Indians,* vol. 6, *Social Anthropology,* edited by Manning Nash, Austin: University of Texas Press.

Wolf, Eric R., and Edward C. Hansen (1972), *The Human Condition in Latin America,* London and Toronto: Oxford University Press.

Womack Jr, John (1969), *Zapata and the Mexican Revolution,* New York: Alfred A. Knopf.

Zavala, Silvio, and Maria Castelo (1939–45), *Fuentes Para la Historia del Trabajo en la Nueva España,* 8vols., Mexico City: Fondo de Cultura Economica.

Zavala, Silvio, and Jose Miranda (1954), 'Instituciones Indígenas en la Colonia', in *Metodos y Resultados de la Politica Indigenista en Mexico,* by Alfonso Caso *et al.*, Memorias del Instituto Nacional Indigenista.

Zorita, Alonso de (1942), *Los Señores de la Nueva España,* Mexico City: U.N.A.M.

INDEX